The Body: A Very Short Introduction

THE BIBLE John Riches
BIBLICAL ARCHAEOLOGY Eric H. Cline
BIOGRAPHY Hermione Lee
BLACK HOLES Katherine Blundell
THE BLUES Elijah Wald
THE BODY Chris Shilling
THE BOOK OF MORMON Terryl Givens
BORDERS Alexander C. Diener and Joshua Hagen
THE BRAIN Michael O'Shea
THE BRITISH CONSTITUTION Martin Loughlin
THE BRITISH EMPIRE Ashley Jackson
BRITISH POLITICS Anthony Wright
BUDDHA Michael Carrithers
BUDDHISM Damien Keown
BUDDHIST ETHICS Damien Keown
BYZANTIUM Peter Sarris
CANCER Nicholas James
CAPITALISM James Fulcher
CATHOLICISM Gerald O'Collins
CAUSATION Stephen Mumford and Rani Lill Anjum
THE CELL Terence Allen and Graham Cowling
THE CELTS Barry Cunliffe
CHAOS Leonard Smith
CHEMISTRY Peter Atkins
CHILD PSYCHOLOGY Usha Goswami
CHILDREN'S LITERATURE Kimberley Reynolds
CHINESE LITERATURE Sabina Knight
CHOICE THEORY Michael Allingham
CHRISTIAN ART Beth Williamson
CHRISTIAN ETHICS D. Stephen Long
CHRISTIANITY Linda Woodhead
CITIZENSHIP Richard Bellamy
CIVIL ENGINEERING David Muir Wood
CLASSICAL LITERATURE William Allan
CLASSICAL MYTHOLOGY Helen Morales
CLASSICS Mary Beard and John Henderson
CLAUSEWITZ Michael Howard
CLIMATE Mark Maslin
THE COLD WAR Robert McMahon
COLONIAL AMERICA Alan Taylor
COLONIAL LATIN AMERICAN LITERATURE Rolena Adorno
COMEDY Matthew Bevis
COMMUNISM Leslie Holmes
COMPLEXITY John H. Holland
THE COMPUTER Darrel Ince
CONFUCIANISM Daniel K. Gardner
THE CONQUISTADORS Matthew Restall and Felipe Fernández-Armesto
CONSCIENCE Paul Strohm
CONSCIOUSNESS Susan Blackmore
CONTEMPORARY ART Julian Stallabrass
CONTEMPORARY FICTION Robert Eaglestone
CONTINENTAL PHILOSOPHY Simon Critchley
CORAL REEFS Charles Sheppard
CORPORATE SOCIAL RESPONSIBILITY Jeremy Moon
CORRUPTION Leslie Holmes
COSMOLOGY Peter Coles
CRIME FICTION Richard Bradford
CRIMINAL JUSTICE Julian V. Roberts
CRITICAL THEORY Stephen Eric Bronner
THE CRUSADES Christopher Tyerman
CRYPTOGRAPHY Fred Piper and Sean Murphy
THE CULTURAL REVOLUTION Richard Curt Kraus
DADA AND SURREALISM David Hopkins
DANTE Peter Hainsworth and David Robey
DARWIN Jonathan Howard
THE DEAD SEA SCROLLS Timothy Lim
DEMOCRACY Bernard Crick
DERRIDA Simon Glendinning
DESCARTES Tom Sorell
DESERTS Nick Middleton
DESIGN John Heskett
DEVELOPMENTAL BIOLOGY Lewis Wolpert
THE DEVIL Darren Oldridge
DIASPORA Kevin Kenny
DICTIONARIES Lynda Mugglestone
DINOSAURS David Norman
DIPLOMACY Joseph M. Siracusa
DOCUMENTARY FILM Patricia Aufderheide
DREAMING J. Allan Hobson
DRUGS Leslie Iversen
DRUIDS Barry Cunliffe
EARLY MUSIC Thomas Forrest Kelly

THE EARTH Martin Redfern
ECONOMICS Partha Dasgupta
EDUCATION Gary Thomas
EGYPTIAN MYTH Geraldine Pinch
EIGHTEENTH-CENTURY BRITAIN Paul Langford
THE ELEMENTS Philip Ball
EMOTION Dylan Evans
EMPIRE Stephen Howe
ENGELS Terrell Carver
ENGINEERING David Blockley
ENGLISH LITERATURE Jonathan Bate
THE ENLIGHTENMENT John Robertson
ENTREPRENEURSHIP Paul Westhead and Mike Wright
ENVIRONMENTAL ECONOMICS Stephen Smith
EPICUREANISM Catherine Wilson
EPIDEMIOLOGY Rodolfo Saracci
ETHICS Simon Blackburn
ETHNOMUSICOLOGY Timothy Rice
THE ETRUSCANS Christopher Smith
THE EUROPEAN UNION John Pinder and Simon Usherwood
EVOLUTION Brian and Deborah Charlesworth
EXISTENTIALISM Thomas Flynn
EXPLORATION Stewart A. Weaver
THE EYE Michael Land
FAMILY LAW Jonathan Herring
FASCISM Kevin Passmore
FASHION Rebecca Arnold
FEMINISM Margaret Walters
FILM Michael Wood
FILM MUSIC Kathryn Kalinak
THE FIRST WORLD WAR Michael Howard
FOLK MUSIC Mark Slobin
FOOD John Krebs
FORENSIC PSYCHOLOGY David Canter
FORENSIC SCIENCE Jim Fraser
FORESTS Jaboury Ghazoul
FOSSILS Keith Thomson
FOUCAULT Gary Gutting
FRACTALS Kenneth Falconer
FREE SPEECH Nigel Warburton
FREE WILL Thomas Pink
FRENCH LITERATURE John D. Lyons
THE FRENCH REVOLUTION William Doyle
FREUD Anthony Storr
FUNDAMENTALISM Malise Ruthven
GALAXIES John Gribbin
GALILEO Stillman Drake
GAME THEORY Ken Binmore
GANDHI Bhikhu Parekh
GENES Jonathan Slack
GENIUS Andrew Robinson
GEOGRAPHY John Matthews and David Herbert
GEOPOLITICS Klaus Dodds
GERMAN LITERATURE Nicholas Boyle
GERMAN PHILOSOPHY Andrew Bowie
GLOBAL CATASTROPHES Bill McGuire
GLOBAL ECONOMIC HISTORY Robert C. Allen
GLOBALIZATION Manfred Steger
GOD John Bowker
THE GOTHIC Nick Groom
GOVERNANCE Mark Bevir
THE GREAT DEPRESSION AND THE NEW DEAL Eric Rauchway
HABERMAS James Gordon Finlayson
HAPPINESS Daniel M. Haybron
HEGEL Peter Singer
HEIDEGGER Michael Inwood
HERMENEUTICS Jens Zimmermann
HERODOTUS Jennifer T. Roberts
HIEROGLYPHS Penelope Wilson
HINDUISM Kim Knott
HISTORY John H. Arnold
THE HISTORY OF ASTRONOMY Michael Hoskin
THE HISTORY OF LIFE Michael Benton
THE HISTORY OF MATHEMATICS Jacqueline Stedall
THE HISTORY OF MEDICINE William Bynum
THE HISTORY OF TIME Leofranc Holford-Strevens
HIV/AIDS Alan Whiteside
HOBBES Richard Tuck
HORMONES Martin Luck
HUMAN ANATOMY Leslie Klenerman
HUMAN EVOLUTION Bernard Wood
HUMAN RIGHTS Andrew Clapham
HUMANISM Stephen Law
HUME A. J. Ayer
HUMOUR Noël Carroll
THE ICE AGE Jamie Woodward

IDEOLOGY Michael Freeden
INDIAN PHILOSOPHY Sue Hamilton
INFECTIOUS DISEASE Marta L. Wayne and Benjamin M. Bolker
INFORMATION Luciano Floridi
INNOVATION Mark Dodgson and David Gann
INTELLIGENCE Ian J. Deary
INTERNATIONAL LAW Vaughan Lowe
INTERNATIONAL MIGRATION Khalid Koser
INTERNATIONAL RELATIONS Paul Wilkinson
INTERNATIONAL SECURITY Christopher S. Browning
IRAN Ali M. Ansari
ISLAM Malise Ruthven
ISLAMIC HISTORY Adam Silverstein
ITALIAN LITERATURE Peter Hainsworth and David Robey
JESUS Richard Bauckham
JOURNALISM Ian Hargreaves
JUDAISM Norman Solomon
JUNG Anthony Stevens
KABBALAH Joseph Dan
KAFKA Ritchie Robertson
KANT Roger Scruton
KEYNES Robert Skidelsky
KIERKEGAARD Patrick Gardiner
KNOWLEDGE Jennifer Nagel
THE KORAN Michael Cook
LANDSCAPE ARCHITECTURE Ian H. Thompson
LANDSCAPES AND GEOMORPHOLOGY Andrew Goudie and Heather Viles
LANGUAGES Stephen R. Anderson
LATE ANTIQUITY Gillian Clark
LAW Raymond Wacks
THE LAWS OF THERMODYNAMICS Peter Atkins
LEADERSHIP Keith Grint
LIBERALISM Michael Freeden
LIGHT Ian Walmsley
LINCOLN Allen C. Guelzo
LINGUISTICS Peter Matthews
LITERARY THEORY Jonathan Culler
LOCKE John Dunn
LOGIC Graham Priest
LOVE Ronald de Sousa
MACHIAVELLI Quentin Skinner
MADNESS Andrew Scull
MAGIC Owen Davies
MAGNA CARTA Nicholas Vincent
MAGNETISM Stephen Blundell
MALTHUS Donald Winch
MANAGEMENT John Hendry
MAO Delia Davin
MARINE BIOLOGY Philip V. Mladenov
THE MARQUIS DE SADE John Phillips
MARTIN LUTHER Scott H. Hendrix
MARTYRDOM Jolyon Mitchell
MARX Peter Singer
MATERIALS Christopher Hall
MATHEMATICS Timothy Gowers
THE MEANING OF LIFE Terry Eagleton
MEDICAL ETHICS Tony Hope
MEDICAL LAW Charles Foster
MEDIEVAL BRITAIN John Gillingham and Ralph A. Griffiths
MEDIEVAL LITERATURE Elaine Treharne
MEMORY Jonathan K. Foster
METAPHYSICS Stephen Mumford
MICHAEL FARADAY Frank A. J. L. James
MICROBIOLOGY Nicholas P. Money
MICROECONOMICS Avinash Dixit
MICROSCOPY Terence Allen
THE MIDDLE AGES Miri Rubin
MINERALS David Vaughan
MODERN ART David Cottington
MODERN CHINA Rana Mitter
MODERN FRANCE Vanessa R. Schwartz
MODERN IRELAND Senia Pašeta
MODERN JAPAN Christopher Goto-Jones
MODERN LATIN AMERICAN LITERATURE Roberto González Echevarría
MODERN WAR Richard English
MODERNISM Christopher Butler
MOLECULES Philip Ball
THE MONGOLS Morris Rossabi
MOONS David A. Rothery
MORMONISM Richard Lyman Bushman
MOUNTAINS Martin F. Price
MUHAMMAD Jonathan A. C. Brown
MULTICULTURALISM Ali Rattansi
MUSIC Nicholas Cook
MYTH Robert A. Segal

THE NAPOLEONIC WARS Mike Rapport
NATIONALISM Steven Grosby
NELSON MANDELA Elleke Boehmer
NEOLIBERALISM Manfred Steger and Ravi Roy
NETWORKS Guido Caldarelli and Michele Catanzaro
THE NEW TESTAMENT Luke Timothy Johnson
THE NEW TESTAMENT AS LITERATURE Kyle Keefer
NEWTON Robert Iliffe
NIETZSCHE Michael Tanner
NINETEENTH-CENTURY BRITAIN Christopher Harvie and H. C. G. Matthew
THE NORMAN CONQUEST George Garnett
NORTH AMERICAN INDIANS Theda Perdue and Michael D. Green
NORTHERN IRELAND Marc Mulholland
NOTHING Frank Close
NUCLEAR PHYSICS Frank Close
NUCLEAR POWER Maxwell Irvine
NUCLEAR WEAPONS Joseph M. Siracusa
NUMBERS Peter M. Higgins
NUTRITION David A. Bender
OBJECTIVITY Stephen Gaukroger
THE OLD TESTAMENT Michael D. Coogan
THE ORCHESTRA D. Kern Holoman
ORGANIZATIONS Mary Jo Hatch
PAGANISM Owen Davies
THE PALESTINIAN-ISRAELI CONFLICT Martin Bunton
PARTICLE PHYSICS Frank Close
PAUL E. P. Sanders
PEACE Oliver P. Richmond
PENTECOSTALISM William K. Kay
THE PERIODIC TABLE Eric R. Scerri
PHILOSOPHY Edward Craig
PHILOSOPHY IN THE ISLAMIC WORLD Peter Adamson
PHILOSOPHY OF LAW Raymond Wacks
PHILOSOPHY OF SCIENCE Samir Okasha
PHOTOGRAPHY Steve Edwards
PHYSICAL CHEMISTRY Peter Atkins
PILGRIMAGE Ian Reader
PLAGUE Paul Slack
PLANETS David A. Rothery
PLANTS Timothy Walker
PLATE TECTONICS Peter Molnar
PLATO Julia Annas
POLITICAL PHILOSOPHY David Miller
POLITICS Kenneth Minogue
POSTCOLONIALISM Robert Young
POSTMODERNISM Christopher Butler
POSTSTRUCTURALISM Catherine Belsey
PREHISTORY Chris Gosden
PRESOCRATIC PHILOSOPHY Catherine Osborne
PRIVACY Raymond Wacks
PROBABILITY John Haigh
PROGRESSIVISM Walter Nugent
PROTESTANTISM Mark A. Noll
PSYCHIATRY Tom Burns
PSYCHOANALYSIS Daniel Pick
PSYCHOLOGY Gillian Butler and Freda McManus
PSYCHOTHERAPY Tom Burns and Eva Burns-Lundgren
PURITANISM Francis J. Bremer
THE QUAKERS Pink Dandelion
QUANTUM THEORY John Polkinghorne
RACISM Ali Rattansi
RADIOACTIVITY Claudio Tuniz
RASTAFARI Ennis B. Edmonds
THE REAGAN REVOLUTION Gil Troy
REALITY Jan Westerhoff
THE REFORMATION Peter Marshall
RELATIVITY Russell Stannard
RELIGION IN AMERICA Timothy Beal
THE RENAISSANCE Jerry Brotton
RENAISSANCE ART Geraldine A. Johnson
REVOLUTIONS Jack A. Goldstone
RHETORIC Richard Toye
RISK Baruch Fischhoff and John Kadvany
RITUAL Barry Stephenson
RIVERS Nick Middleton
ROBOTICS Alan Winfield
ROMAN BRITAIN Peter Salway
THE ROMAN EMPIRE Christopher Kelly
THE ROMAN REPUBLIC David M. Gwynn
ROMANTICISM Michael Ferber
ROUSSEAU Robert Wokler
RUSSELL A. C. Grayling
RUSSIAN HISTORY Geoffrey Hosking

RUSSIAN LITERATURE Catriona Kelly
THE RUSSIAN REVOLUTION S. A. Smith
SCHIZOPHRENIA Chris Frith and Eve Johnstone
SCHOPENHAUER Christopher Janaway
SCIENCE AND RELIGION Thomas Dixon
SCIENCE FICTION David Seed
THE SCIENTIFIC REVOLUTION Lawrence M. Principe
SCOTLAND Rab Houston
SEXUALITY Véronique Mottier
SIKHISM Eleanor Nesbitt
THE SILK ROAD James A. Millward
SLEEP Steven W. Lockley and Russell G. Foster
SOCIAL AND CULTURAL ANTHROPOLOGY John Monaghan and Peter Just
SOCIAL PSYCHOLOGY Richard J. Crisp
SOCIAL WORK Sally Holland and Jonathan Scourfield
SOCIALISM Michael Newman
SOCIOLINGUISTICS John Edwards
SOCIOLOGY Steve Bruce
SOCRATES C. C. W. Taylor
SOUND Mike Goldsmith
THE SOVIET UNION Stephen Lovell
THE SPANISH CIVIL WAR Helen Graham
SPANISH LITERATURE Jo Labanyi
SPINOZA Roger Scruton
SPIRITUALITY Philip Sheldrake
SPORT Mike Cronin
STARS Andrew King
STATISTICS David J. Hand
STEM CELLS Jonathan Slack
STRUCTURAL ENGINEERING David Blockley
STUART BRITAIN John Morrill
SUPERCONDUCTIVITY Stephen Blundell
SYMMETRY Ian Stewart
TAXATION Stephen Smith
TEETH Peter S. Ungar
TERRORISM Charles Townshend
THEATRE Marvin Carlson
THEOLOGY David F. Ford
THOMAS AQUINAS Fergus Kerr
THOUGHT Tim Bayne
TIBETAN BUDDHISM Matthew T. Kapstein
TOCQUEVILLE Harvey C. Mansfield
TRAGEDY Adrian Poole
THE TROJAN WAR Eric H. Cline
TRUST Katherine Hawley
THE TUDORS John Guy
TWENTIETH-CENTURY BRITAIN Kenneth O. Morgan
THE UNITED NATIONS Jussi M. Hanhimäki
THE U.S. CONGRESS Donald A. Ritchie
THE U.S. SUPREME COURT Linda Greenhouse
UTOPIANISM Lyman Tower Sargent
THE VIKINGS Julian Richards
VIRUSES Dorothy H. Crawford
WATER John Finney
WILLIAM SHAKESPEARE Stanley Wells
WITCHCRAFT Malcolm Gaskill
WITTGENSTEIN A. C. Grayling
WORK Stephen Fineman
WORLD MUSIC Philip Bohlman
THE WORLD TRADE ORGANIZATION Amrita Narlikar
WORLD WAR II Gerhard L. Weinberg
WRITING AND SCRIPT Andrew Robinson

Available soon:

GOETHE Ritchie Robertson
FUNGI Nicholas P. Money
ENVIRONMENTAL POLITICS Andrew Dobson
MODERN DRAMA Kirsten E. Shepherd-Barr
THE MEXICAN REVOLUTION Alan Knight

For more information visit our website

www.oup.com/vsi/

Chris Shilling

THE BODY

A Very Short Introduction

OXFORD
UNIVERSITY PRESS

Great Clarendon Street, Oxford, OX2 6DP,
United Kingdom

Oxford University Press is a department of the University of Oxford.
It furthers the University's objective of excellence in research, scholarship,
and education by publishing worldwide. Oxford is a registered trade mark of
Oxford University Press in the UK and in certain other countries

First edition published in 2016

Published in the United States of America by Oxford University Press
198 Madison Avenue, New York, NY 10016, United States of America

British Library Cataloguing in Publication Data
Data available

Library of Congress Control Number: 2015945425

ISBN 978-0-19-873903-6

Printed and bound by
CPI Group (UK) Ltd, Croydon, CR0 4YY

Contents

Preface xiii

Acknowledgements xvii

List of illustrations xix

Introduction 1

1 Natural bodies or social bodies? 7

2 Sexed bodies 24

3 Educating bodies 42

4 Governing bodies 60

5 Bodies as commodities 80

6 Bodies matter: dilemmas and controversies 97

References 109

Further reading 115

Index 117

Preface

I first developed an academic interest in bodies when I was a postgraduate student completing my PhD in the mid-1980s. My thesis focused upon school-based vocational schemes that sought to equip students with those skills and attitudes that governments had suggested would enhance their 'employability'. The longer I spent conducting research in educational institutions, training courses, and work-experience placements, however, the more perplexed I became at most of the literature published on the subject.

In books and articles on schools, for example, individuals were portrayed as shadowy carriers of 'linguistic codes', as ciphers of social class forces beyond their control, and as variably competent cognitive processors of educational knowledge. Despite the noise of morning bells and shouting teachers, the verbal protests and physical jostling of students rushing along brightly lit corridors, and the pungent smells escaping school canteens and gyms, no one seemed to possess a living, sensing *body*. More often than not, this 'disembodied' approach was replicated in analyses of other forms of education and training.

Educational knowledge was discussed at length in these writings, but mostly in terms of academic qualifications, abstract propositions about truth, the ideological effects of the curriculum, or how acquiring skills was associated with the reproduction of social class inequalities. It was rare to hear much about education's role in imparting to students particular physical experiences, habits, disciplines, and practical techniques. Also conspicuous by their infrequency were convincing attempts to evoke the tactile challenges of working with contrasting materials, or to map the novice's attempts to 'get to grips' with stubborn physical equipment that failed to bend to their will. Even more lacking was a sense of how thought and knowledge, of *whatever* variety, was engaged in and acquired by humans who, as embodied organisms, were always *physically* located in and engaged with the environments in which they lived, worked, and rested.

These analyses were not alone in their approach towards the body, but followed and drew on a long tradition of philosophical and theological inquiry in the West. Yet the more I read around the subject after completing my PhD, returning to my interest in social and political thought during my first years lecturing at Oxford Polytechnic and then Southampton University, the more it became clear that there existed a range of writings across philosophy, sociology, religion, and history that could be interpreted as contributing towards an interdisciplinary field of what I later referred to as 'body studies'.

Since that time, I have been fascinated by how the subject of human embodiment can provide us with a productive starting point from which to explore a wide range of issues central to the social sciences, humanities, and arts, and have been fortunate enough to write and teach widely in and around the subject. Focusing on embodiment makes it possible to connect the personal with the political, the intellectual with the practical, the symbolic with the sensual, and to appreciate the potential for shared experience and understanding that underpins the social, ethnic, religious, and cultural differences that drive so much contemporary global conflict. At the same time, it highlights the fundamental differences that separate people whose senses, feelings, and actions have been subjected to radically different forms of embodied pedagogy since birth, and the problematic nature of theories that suggest 'rational dialogue' or 'ideal speech' situations are by themselves able to resolve conflict.

The virtually unlimited range of issues raised by embodiment, and the thoroughly interdisciplinary character of body studies, presents a particular challenge when it comes to writing a *Very Short Introduction*. It is impossible to be comprehensive and there are a number of important issues that I have had to omit in order to remain within the constraints of this format. Nevertheless, I have tried to convey a sense of the exciting scope of this field and to highlight its significance for understanding what is going on in the world today. In so doing, I have also organized my discussion around several key themes—set out in the Introduction—that inform each chapter and are designed to provide coherence to the range of body matters discussed throughout this book. My hope is that this has resulted in a treatment of embodiment that will be of interest to the lay reader and that will also provide a clear and easily accessible 'updating' of the body's significance to those who already have some basic knowledge of the field.

Acknowledgements

This may be a *Very Short Introduction*, but in writing it I have accumulated a number of debts. I would like to thank Andrea Keegan and Jenny Nugee at Oxford University Press for their encouragement and efficiency, Larry Ray and Vince Miller for commenting on individual chapters, and the external readers for their detailed engagement with the manuscript. Many thanks also to Philip A. Mellor for his advice, good sense, and comradeship, and to Leif Östman and colleagues on the Embodied Knowledge project based at the University of Uppsala for their warm collegiality. More generally, Sarah Vickerstaff and the rest of my colleagues at the University of Kent have provided a supportive environment in which to work, while I owe much to Kalli Glezakou and her team in the PG office. Closer to home, or rather at home, Debbie, Max, and Kate were good enough to provide feedback on and suggestions for various chapters, and so much more. This book is dedicated to them.

List of illustrations

1 Female bodybuilder **14**
© Maria Rutherford/Getty Images

2 Tattooed individual **15**
Guy Corbishley/Getty Images

3 Performing femininity **37**
Ray Kachatorian/Getty Images

4 Transgendered individual **39**
Patryce Bak/Getty Images

5 Sports in school **44**
Peter Cade/Getty Images

6 The hanging of Haman **63**
DEA/G. DAGLI ORTI/Getty Images

7 Bentham's panopticon **65**
Photo by Mansell/The LIFE Picture Collection/Getty Images

8 Production lines can facilitate the surveillance of workers **67**
© Cultura Creative (RF)/Alamy

9 The commodification of sex **93**
© HBimages/Alamy

10 Dress is, for many, an integral aspect of religious identity **105**
© Pilchards/Alamy

Introduction

Body matters are rarely out of the news, occupying as they do a prominent place in the concerns of politicians, scientists, health experts, educators, moralists, and religious authorities. Whether the focus is on national sporting performance and the 'obesity epidemic', global inequalities in morbidity and mortality rates, 'alien' modes of religious dress, or the latest technological means of supplementing our capacities through prostheses, digital media, and neural implants, issues surrounding human embodiment are frequently subject to debate and disagreement.

The variety of interests and agendas associated with such controversies suggests that understanding the meaning and social significance of embodiment requires us to travel well beyond the parameters of physiology or other branches of the biological sciences. This task has increasingly informed the agendas of the social sciences and humanities over the last three decades, and it is the aim of this book to make sense of the various ways in which the body has been analysed during that time while also exploring some of the most important body matters facing society today.

These comments do not mean to imply that the body has only recently become a subject of social and cultural importance. In Ancient Greece, for example, philosophers devoted sustained attention to the relationship between the body and the mind, while artistic representations of physical perfection from that era continue to influence contemporary conceptions of the 'body beautiful'. More generally, forms of body modification such as tattooing, scarification, and cosmetic surgery not only have a long history, but also raised in the past social issues regarding ethnicity, class, and gender that remain relevant and contentious today.

The development and popularity of particular cosmetic surgery procedures, for example, has been informed deeply by the existence of power inequalities between different groups. In the late 19th century, John Roe developed intranasal operations designed to correct the 'pug nose', a characteristic stigmatized through its association with lower class Irish immigrants. In the early decades of the 20th century, another pioneer of modern rhinoplasty, Jacques Joseph, developed and performed procedures on German Jews that helped them to become 'ethnically invisible' facially. Following World War II, 'double eyelid surgery' grew in popularity among Chinese, Korean, Japanese, and Asian Americans wishing to look more Western, while breast reduction in Rio de Janeiro has been linked to a desire to avoid the association of 'pendulous breasts' with the black working classes (an image associated with slavery). In each of these cases body matters highlight how the physical self can be changed, but also connect us to a range of social and historical issues.

If the body is 'good to think with' it has not always been seen as a central part of what makes us human or social beings. In contrast to various Eastern traditions of thought and practice, such as Confucianism and Daoism, the dominant Western approach to philosophy has tended to relegate its significance. Few were more influential in this respect than the 17th-century philosopher Rene Descartes. Famous for his dictum 'I think therefore I am', Descartes assumed a strong distinction between the mind, on the one hand, and the body's senses, on the other, and prioritized the former over the latter for his assessment of what it meant to be human. Descartes' thought can itself be located within a tradition of Judeo-Christian visions of individuals as 'dualistic beings' irrevocably split between mind/soul and body. Yet his focus on the isolated mind as a generator of ideas—encased within the individual and separated from the 'external' world—was not the only approach on offer.

In what has been referred to as a 'marginalized history' of body relevant writings in philosophy, for example, several contributions have been identified as particularly important. Friedrich Nietzsche emphasized how Western values entailed the sublimation of intoxication, sexuality, and violence—which are experiences linked directly with bodily feelings and expressions. Maurice Merleau-Ponty identified the body as our vehicle in and vantage point upon the world, with our senses 'unfolding' onto their surroundings. John Dewey analysed cycles of habit, crisis, and creativity as informing the relationship of embodied subjects to their environment. More recently, the controversial French philosopher Michel Foucault sought to show how distinctive forms of knowledge and power have historically exerted a pervasive effect on people's bodies.

The importance of the body goes beyond abstract philosophical formulations, however, irrespective of how provocative or impressive these may be. Our embodied existence can be seen more generally as a foundation from which can be built an empirically informed yet distinctive approach to the analysis of society, identity, culture, and history. At its most basic, this involves recognizing that people's ability to make a difference to their own lives, and to those of others, depends on them being, having, and using their bodies in order to intervene in the 'flow' of social life. Relatedly, the capacity of governmental and other authorities to direct our actions, and manage populations, depends on them being able to gather knowledge about and exert control over our physical movements. Bodies, in short, are important practical as well as intellectual matters.

In seeking to develop and highlight the utility of such an approach, I focus in this book on a limited number of body matters that enable me to demonstrate clearly the importance of an embodied perspective on the world in which we live. After exploring the rise and parameters of this interdisciplinary area of study (an area associated with a certain convergence between the social sciences and the biological sciences), I devote further chapters to 'sexed bodies', 'governed bodies', 'educated bodies', and 'bodies as commodities'. Each of these topics involves a range of contrasting issues, but they also share in common three themes that inform and provide direction to the main arguments in this book and become the explicit focus of the concluding chapter.

The first of these themes concerns the significance of social and technological forces for the constitution and development of what is frequently understood to be the biological constitution of our embodied being. Social variables such as inequalities in earnings have been shown to affect levels of illness and life expectancy, for example, while science, technology, and medicine have demonstrated that the body and brain (itself an integral part of our embodied being) can be investigated, managed, and changed in a number of different ways. Against this background it has become increasingly evident that society, in the broadest sense of that term, influences our physical being at the most profound levels, and that it is difficult to disentangle the social from the biological processes that affect what and who we are.

The second theme emerges directly from recognizing that society is significant for our bodily being. If social, technological, and medical forces make it possible to exert increasing control over bodies, undermining the idea that they are unalterable biological organisms, this raises questions and dilemmas about *how* we should manage and control our bodily being. Transplant surgery, *in vitro* fertilization, weight-loss surgery, genetic engineering, and the prospect that DNA testing kits presage a new era of personalized medicine suggest that the body is, for the affluent, becoming a matter of *options* and *choices*. The technical ability to change the body, however, often exceeds existing moral frameworks that prescribe what is 'natural' about the body. Such developments even raise questions about what the body is and can do, casting doubt on traditional assumptions regarding the limitations and capacities of embodied subjects.

Amid these doubts and alternatives many people have become increasingly *reflexive* about their physical selves. What I mean by this is that the number of opportunities for embodied change that exist in the current era often encourages individuals to think about their previous routines and actions, physical competences and appearances, rather than just to continue accepting them and living habitually. These thoughts or conscious reflections involve not only assessing the past state and capacities of our bodily being, but also projecting into the future possible versions of ourselves reformed on the basis of new commitments. Acknowledging the importance of such reflexive thinking does not suggest that mental thoughts are somehow separate from bodily being. Rather, it recognizes that traditional physical habits and ways of knowing the body are subject to increasing levels of reassessment and reappraisal in the light of current developments.

The third theme informing the later chapters in the book focuses on how particular views of, or approaches to, the body attribute value to people's physical selves in very different ways. Since the origins of slavery people's bodies have been prized as commodities, as resources able to create wealth for their owners. The current significance of human trafficking, forced labour, and the sex trade suggests that this view of the body's value remains globally significant. Governmental attempts to cultivate the body via education and training, in contrast, may seek to add value to embodied subjects by enhancing their capacities to engage in particular activities and pursue specific goals. Contemporarily, indeed, there has been a proliferation of ways in which bodies have become valuable, prized, and even sacred. These range from conceptions of the body as esteemed because of its capacity to act as a vehicle for profit, to religious views of divinely sanctioned forms of prayer, dress, diet, and other bodily practices.

I want to conclude this Introduction with a note on terminology. In writing about 'bodies' I use this term mostly as shorthand for the embodied human *as a whole*. It serves to highlight the often academically neglected physical and organic dimensions of existence, and how these are centrally significant to people's identities, actions, and relationships. 'The body' also at times in what follows refers in a more restricted sense to the physical flesh, as reflected upon by embodied individuals when they are thinking or talking about such issues as their appearance, and I make it clear when this is the case. Irrespective of the specific terminology used to talk about body matters throughout this book, however, my aim is to demonstrate the necessity of understanding humans as embodied beings if we are to advance our understanding of the social and material world in which we live and on which we depend.

Chapter 1
Natural bodies or social bodies?

One of the main themes informing this book entails the need to demonstrate how social factors are important for the constitution and development of our embodied being. Yet not everyone will be convinced that this is a feasible undertaking. The human body is still considered by some to be an exclusively biological entity. From this perspective, our longevity, morbidity, size, and appearance is determined by genetic factors—themselves the product of long-term evolutionary processes—that render insignificant the influence of society.

Since the 1980s, however, there has emerged from the social sciences and humanities a broad interdisciplinary area of research and writing known as 'body studies'. This academic field addresses a wide variety of social and cultural issues viewed as relevant to the inescapably bodily character of human existence. These range from how traditional societies maintained collective identities among their members by tattooing and scarifying the flesh of initiates during religious rites, to historical changes in beliefs about the constitution and limitations of men's and women's bodies. They further include such matters as the impact diet and work have on people's health, and the widespread significance of the body as a 'natural symbol' that helps people to think, classify, and even engage in discriminatory practices based on physical markers associated with ethnicity, 'race', age, sex, and disability.

Diverse though these issues are, they each reveal how the physical appearances, capacities, and experiences of humans bear the imprint of socially and culturally specific ways of living in and engaging with the wider environment. They also highlight the significance of the body for a range of academic disciplines regarded conventionally as possessing both subject matter and methods very different from those associated with the biological sciences.

If we want to understand how the body became a viable and popular subject for the social sciences and humanities, we need to explore the context in which this occurred. This included a number of social and historical developments—involving, for example, feminist and environmental campaigns, medical and technological developments, and the rise of consumer culture—that raised the visibility of the body as a general academic issue. Also relevant was the readiness of certain social scientists and biological scientists to reflect on, expand, and seek to enhance their own analyses by taking seriously a subject matter they had previously ignored.

Why the body? Social factors

A number of social and historical factors prepared the ground for the growing interdisciplinary interest in the body as an academic issue during the 1980s and 1990s. Each highlighted different aspects of the body's importance for society, increasing the scope of what could count as body matters for academics. Six of them were especially influential.

First, the resurgence of 'second wave' feminism in the 1960s and 1970s made political those personal bodily issues related to inequalities in health provision, abortion, rape, pornography, and prostitution that were harmful to the interests of women. The 'reclaim the night' marches during the 1970s, for example, protested against the physical and sexual violence women risked by simply being visible in public places at times when society expected them to be indoors, fulfilling their 'authorized' roles of wife, mother, or daughter. In this case, as in many others related to sexual, 'racial', and other inequalities explored in this book, the body provided a link between the restrictions and risks felt by individuals and the wider social position of particular groups.

Second, the growth of political radicalism, 'alternative lifestyles', and ecological concerns in North America and Europe during this same period highlighted in a distinctive way the significance of the relationship between the human body and key dimensions of the wider environment. Left-wing critics and the anti-Vietnam War movement condemned consumer culture, military conflict, and the arms race in part because they were seen as reducing human life to a 'one dimensional' pursuit of wealth and domination. The accusation here was that both embodied individuals and the natural environment were being treated by military-industrial societies as disposable means towards narrow, instrumental, and damaging ends.

The concerns about environmental sustainability evident in these criticisms were given added impetus by the Club of Rome's 1972 report *Limits to Growth.* This global think tank highlighted the risks to future life on earth by exploring trends in population growth, food production, pollution, and the industrial consumption of non-renewable natural resources. Since then, problems associated with climate change, soil erosion and deforestation, nuclear waste, the exhaustion of fossil fuels, and food shortages have been at the forefront of the Green movement's warnings that the planet is being treated as a mere resource for productivity, raising doubts about its long-term capacity to sustain the embodied life of humans or other animals.

The third factor that stimulated academic interest in the body involved the 'ageing' of societies in the Global North, as well as in areas of the Global South. Medical advances together with the spread and improvement of basic amenities have raised life expectancy for many, and the United Nations predicts that there is likely to be a doubling of the percentage of the world's population over the age of 60 (from 10 per cent to 21 per cent) between 2000 and 2050. Living further into old age, however, often means having to live more years of life with chronic health problems. This has become an increasingly important political issue since the 2007 financial crisis in which the 'burden' of caring for increasing numbers of aged people, coupled with high levels of youth unemployment in Europe and other regions, has been associated with the possibility of generational conflict over the distribution of resources.

Rising numbers of ageing bodies have also stimulated interest in the relationship between culture and embodiment given that the very process of growing old, and becoming dependent, has been stigmatized in the most economically influential regions of the world. Anglo-American films, media, and advertising prize the slim, sexy, young, and *independent* body. At the same time, those who fall outside of this idealized image are presented with the opportunity to 'save themselves' and become 'reborn' by reflecting on their limitations and embarking upon cosmetic, exercise, and surgical options necessary to defy the appearance of age and pursue the prize of 'eternal youth' and personal autonomy. Those who refuse to or are unable to succeed in this challenge of maintaining their bodies in a socially valued form risk accusations of becoming a burden on society (one of the concerns raised by opponents of euthanasia). Once again, it is bodies that become the medium through which social relations are constructed.

The fourth factor to have increased the social visibility of the body is related to a structural change within advanced capitalism during the second half of the 20th century. In place of the 'save and invest' mentality promoted by governments seeking to stimulate economic growth in earlier decades, the development of consumer culture during this period encouraged individuals to achieve status by purchasing material goods. This was associated with the body becoming an 'object for display' in media and advertising through commercial representations focused initially on women but spreading to men through such cultural inventions as the 'metrosexual' male in the 1990s. Bodies have now become a ubiquitous means through which products are sold and esteem acquired: they have acquired increased value within capitalist societies both as a means of enhancing profitability for producers and as potential status symbols for many consumers.

In certain respects these changes echo trends in the 19th century, a period in which appearance began to be seen as a malleable manifestation of personality rather than as a fixed marker of social position. During this period more individuals began to feel responsible for how they presented themselves in public (before that, public responsibility for self-presentation fell on an aristocratic elite with others often subject to sumptuary laws regarding what they could and could not wear), yet a correlate of this was that people also had to deal with the concern that bodily slips, failures, and embarrassments would be seen as personal failings. The importance of this new emphasis on the body as personality grew during the 20th century and was exemplified by the writings of the American sociologist Erving Goffman. Comparing social life to being 'on stage' during a theatrical performance, Goffman explored how the 'presentation of self' required careful reflection on and management of appearance, behaviour, and speech when engaged in social interaction. This 'impression management' was key not only to the effective performance of social roles, moreover, but also to being seen as possessing a morally acceptable self-identity. Contravening the unspoken rules of interaction or appearance, in contrast, could result in being stigmatized; a fate that those with disabilities often struggled to avoid as a consequence of the prejudices of others. Bodies were not treated equally in the extent to which they could or did exist as markers of social value.

Fifth, there has been an intensified scrutiny of 'alien' bodies following the 9/11 attack on the Twin Towers in America and the subsequent declaration of a 'War on Terror' by the Bush government. These events raised existing concerns about illegal immigration and the threats posed by suicide bombers and terrorists in general. America and other states responded by increasing the use of passport control; identity cards; iris, fingerprint, and voice recognition devices; and the gathering of other biometric data. Here, bodies became scrutinized in terms of the images and data that could be extracted from them as part of attempts to prevent unauthorized persons from entering nation-states. The surveillance of bodies legitimately existing within the spaces occupied by nation-states has also become more common. Covert monitoring of physical movement, telephone conversations, and the traces of our bodily lives left by new social media, Internet activity, and credit card purchases has become a core feature of ostensibly democratic societies.

The sixth major reason prompting a general increase of academic interest in the body is those scientific and technological changes that have facilitated an unprecedented degree of control over the shaping and reshaping of bodies. From the 19th-century emergence of a literature on nutrition, to the 'scientific management' of productivity that reached its height in the early 20th century, to contemporary advances in cosmetic and transplant surgery, our capacity to intervene in the body has never been greater. If such developments provide the potential for achieving heightened levels of control over bodies, however, they have also weakened the boundaries between bodies and technology, reducing our certainty regarding what is 'natural' about the body. It is this uncertainty, and the responses to it evident in the actions of individuals, that provided an additional impetus for the growth of body studies.

Body modification directed towards personal control and transformation is not, of course, new. Early Christians engaged in strict regimes of physical discipline designed to strip them of previous habits prior to being 'reborn', body and soul, through baptism. Cosmetic surgery can be traced back even further, with accounts of facial reconstruction on the living to be found in ancient Indian Sanskrit texts. Bodily alteration has, however, become more extensive and individualized, with fast-changing criteria regarding physical desirability or normality encouraging people to become increasingly reflexive about how they manage various dimensions of their organic selves. In the affluent West, in particular, these developments have led to the body being viewed and treated as a *project*, a 'raw material' to be worked upon as an integral expression of individual self-identity (see Figure 1).

This individualization of bodily self-identity is perhaps exemplified by the changing status of tattooing. Viewed in traditional societies as a marker of collective membership, tattooing has increasingly become a means through which people express their own *individual* sense of selfhood. Creating unique and personal designs, in conjunction with tattoo artists, and using these at times to commemorate major life allegiances or events (e.g. by imprinting their skin with the names of dead loved ones, or by using the ashes of loved ones in the ink used in these images), tattoos have become a means of expressing and stabilizing significant elements of selfhood amid a world in which fashions appear to be changing constantly (see Figure 2).

Other common body projects can be seen in the case of health and fitness (where attempting to *look* healthy has for many become the equivalent of seeking to *become* healthy), dieting, and cosmetic surgery. In 2013 in the USA alone, for example, over $12 billion was spent on surgical and non-surgical cosmetic procedures. Liposuction and breast augmentation were the most popular (with 363,912 and 313,327 operations, respectively), with buttock augmentation and labiaplasty being the fastest growing procedures (the latter in particular illustrating that there are now few areas of the flesh on which people are not prepared to contemplate working).

1. Women bodybuilders provide us with illustrations of how body projects can challenge conventional gender norms.

2. Does tattooing provide a stable means of 'writing identity' onto the self?

The resurgence of religious modes of dress, diet, and other aspects of body management in many parts of the world may appear to qualify this picture of individualized body projects. Yet numerous academic studies, media reports, and blogs suggest that many individuals are *choosing* to continue with or convert to these lifestyles knowing full well (as a result of social media, satellite television, and other channels of global communication) that there exist significant alternatives. What may once have been followed as a result of habit and with reference to tradition is now increasingly pursued through personal reflection, the weighing of alternatives, and the assertion of individual choices centred upon the management and control of the embodied self.

These six social, cultural, and technological developments have highlighted the significance of the body to academics, and their variety helps account for why this field of study is so diverse. Each has focused on different dimensions of the body, assessing as valuable very different elements of embodied existence. Yet these processes alone do not enable us to account fully for why so many scholars from the social sciences and humanities engaged with matters traditionally viewed as the province of the biological sciences. To understand this requires us to appreciate how certain changes in both the social sciences and the biological sciences have prepared the ground for a limited convergence of interest around the body as a cultural as well as an organic phenomenon.

Why the body? Academic factors

The social scientist Ted Benton argued in a series of publications during the 1990s that sociology and the social sciences could only understand adequately people's actions, identities, and relationships if they explored how these were informed by our physical constitution and ecological surroundings. Biology mattered to individuals and to society given that the evolution of advanced primates provided different physical and neurological bases on which social relations could be established, which in turn provided part of the context in which adaptations to the environment and future evolutionary developments occurred. The development of bipedalism and associated advances in tool use and tribal hunting practices, for example, shaped the body in general and the hand in particular. Such changes did much to stimulate developments in the brain, enabling people to advance their skills and specialize in particular tasks, and provided the platform for increasingly complex relations to develop between people.

Thus, while physical and neurological changes might seem initially far removed from the concerns of those interested in social issues, they actually provide an embodied context for modern culture. It is only because we are able to stand upright and develop a manual dexterity guided by our mental powers of reflection and judgement, for example, that we have been able to develop the arts and technologies characteristic of contemporary life. Our bodies, moreover, provide us with the means to interact with other people while also placing constraints upon how this interaction occurs. Our physical interactions with others are structured around the ability to take turns in communicating with others and depend upon people accepting the vulnerability that comes with 'opening themselves up' to others. These interactions are, in other words, *inter*-corporeal—assembled through our mutual corporeality—and even 'virtual' communications facilitated by the Internet and new social media are dependent on our motor capacities and sensory ability to utilize these technologies. Finally, our bodies also inform the metaphors we use to imagine, understand, and classify the world around us: the idea of being 'sick' of someone, of feeling emotionally 'empty', and of not being able to be fully in two places at once are all related to somatic states and experiences.

The importance of organic biological factors to the social aspects of people's identities and capacities is relevant not only to the human species in general, but is also evident over the course of an individual's life. As the body develops from birth and infancy, through puberty and adolescence, and into middle and old age, it provides individuals with contrasting foundations on which to interact with others. The health and strength of our vital organs and senses are inevitably affected by processes of ageing—despite the many ways in which they can be technologically enhanced through spectacles, hearing aids, and other devices—and such factors can impact on an individual's sense of self-identity and their dependency on others. In the UK, for example, approximately 35 per cent of those over 80 years old suffer from serious sight loss; with macular degeneration and other conditions necessitating significant changes to lifestyle. To ignore such factors would be to overlook biological issues that exert an inevitable impact on the autonomy of individuals to act and live as they might wish.

Benton's argument about the biological sciences is not just intended to suggest that social science will remain limited until it takes into account the significance of our physical existence for social life. More positively, it also highlights how social studies can add value to their scope and analytical capacity by exploring an academic terrain they usually avoid. This terrain is well illustrated in the case of bioarchaeology, the study of biological remains from archaeological sites.

In revealing details about the lives of individuals from skeletal remains, bioarchaeology has extended our knowledge of such social matters as gender inequalities, migration, and the effects of political oppression and conflict. Scientists working in this field have shown how the chemical composition of bones reveals major differences in past female and male diets, for example, and can throw fresh light on patterns of geographical mobility. As the bioarchaeologist Rebecca Gowland and the forensic anthropologist Tim Thompson demonstrate, research has also revealed how skeletal remains in slave cemeteries in the Caribbean and North America evidence trauma associated with ill-treatment, poor diet and living conditions, and hard labour. More impressive still is the capacity of bioarchaeology to show how political events—such as the Dutch famine of 1944–5, a period in which Germany stopped food supplies to the Netherlands—affect the health of those developing *in utero*. Findings such as these provide compelling reasons why a growing number of social scientists have come to engage with those dimensions of the biological sciences possessed of the potential to demonstrate the importance of political, cultural, and social relationships and events.

It was not just movement from within the social sciences that prepared the ground for a limited convergence of interest around issues related to human embodiment. Recent developments in the biological sciences have promoted a dynamic and fluid view of human evolution, development, genetics, and the brain. This has increased the sensitivity of these sciences not only to interactions that occur *within* the organism, but also to those that exist *between* the human organism and its social environment.

This sensitivity is manifest in several ways. To begin with, simplistic views of evolution based upon 'the selfish gene'—the idea that human behaviour is controlled by a genetic make-up geared exclusively to ensuring the 'survival of the fittest'—have been contested by more nuanced accounts that provide space for altruistic and other pro-social types of behaviour. These alternative scientific accounts have prompted cultural theorists, such as the feminist writer Elizabeth Grosz, to return to Darwin's original writings in highlighting how processes related to sexual attraction, for example, provide the basis for a sphere of culture relatively independent from issues of survival. In these cases, social and cultural standards become important to social behaviour and human reproduction; a recognition that has enabled analysts to enhance previously rather one-dimensional accounts of what motivates embodied action and interaction.

Relatedly, the field of epigenetics (the study of how changes in gene expression occur) has demonstrated the importance of social and environmental factors to the regulation of genes. Diseases and patterns of physical growth and obesity, for example, are now commonly traced not to the workings of single genes, but to multiple conditions that include interactions between individuals and their social and material surroundings. Epigenetic findings related to the Dutch famine mentioned earlier in this chapter, for example, suggest that political actions resulting in food restriction can affect the second generation of offspring (eggs develop in mothers of the future while they exist as foetuses in mothers of the present). Once more, biological and social processes are revealed to be inextricably related rather than separate phenomena: the socially determined restriction of food to one generation can, remarkably, affect patterns of morbidity and mortality among the grandchildren of those affected.

This dynamic view of how our organic being is open to social influence is shared by cutting edge thinking in neuroscience. Here, the body is viewed as the foundation for the mind: the body's multiple physical receptors (including the retina, cochlea, and the nerve terminals in our skin) receive stimuli from the environment that is turned into a chain of signals that travel to the brain. These signals help construct neural patterns that 'map' our interactions with objects and people in our environment, but sophisticated neurological analyses acknowledge that our brains not only shape but are also *shaped by* these encounters. The brain, it is now widely recognized, is possessed of a plasticity and malleability that is responsive to and develops as a consequence of our actions and interactions.

Recent advances in genetic engineering reinforce further the significance of social factors to our biological constitution: we are now able to alter elements of our evolutionary inheritance. Stem cell research, for example, has made strides that promise to be of enormous significance for medicine, health care, and bioengineering, possessing the potential to produce every type of cell and tissue in the body. The life science industries to which this research is allied began as early as the 1970s to create new products from existing samples of human, animal, plant, and other material by extracting and recombining genetic material in new ways. Such developments provide more evidence of the significance of social developments including, crucially, scientific advances for our biological constitution.

Convergences

If there are compelling reasons for the social sciences to take account of their biological counterparts, it seems that sections of the biological sciences have themselves adopted a more fluid view of their subject matter that is open to the influence of social processes and cultural developments. These movements have resulted in a limited convergence between these two sides of the academic divide around the social importance of the organic body—a recognition that can enhance the value of explanations emanating from a wide range of disciplines. Not everyone holds that social and biological processes are interdependent, but this view is supported by the elementary features of various basic, but socially vital, human capacities such as language and the emotions.

Language development among individuals—so crucial to the cooperation that makes culture and economic prosperity possible—requires not only the tongue, vocal cords, larynx, and other biological equipment necessary for speech, but also the activation and cultivation provided by the social group to which an individual belongs. Emotions—so crucial to the forging of intimate relations—while being essential to the human capacity to detect and respond to danger, also exhibit social and biological dimensions. The fight or flight response, for example, stimulated by a situation in which the organism perceives a threat, *automatically* prepares the individual for action through an increase in adrenaline, raised blood pressure and heart rate, and heightened awareness. The meanings attached to such a response, the feelings experienced by the individual undergoing it, and the actions they engage in during and following it, however, *vary culturally* according to such factors as whether the individual concerned has been exposed to masculine or feminine types of socialization.

If this concern with the social and biological dimensions of human existence has in recent decades highlighted the potential of body studies as an interdisciplinary field, it would be wrong to overlook previous attempts to lay the ground for such a development. Auguste Comte, the founder of sociology, was regarded as one of the leading theorists of biology in 19th-century France, and drew strong parallels between the individual organism and the social organism. Elsewhere, John Dewey, George Herbert Mead, and other pragmatist philosophers of the early 20th century analysed how our physical habits developed as a means of facilitating our survival, possessed a structural basis in our nervous system, and guided our attentiveness and thoughts. William James was another important figure here: possessing a medical degree and teaching physiology at Harvard, he was especially interested in experience. Having insisted on the importance of their organic foundations, however, these writers also argued that our habits are shaped by social relationships and can be scrutinized and changed by an individual's reflections and actions.

The elusiveness of bodies

The field of body studies grew as a result of various developments that highlighted the enfleshment of human existence as a social and political issue, and because of a limited convergence in the interests of social and biological scientists. If the variety of these factors has helped stimulate a new interdisciplinary field of studies, however, it has done little to stabilize answers to the question, 'What is the body?' The more the body is studied, indeed, the more malleable and elusive it seems to become.

The meaning, manifestation, and value attributed to the body has shifted and undergone a metamorphosis depending upon who is studying it. It was a vehicle of domination and oppression for feminists concerned to highlight how women's bodies were being subjugated within society. It was a metric against which to scrutinize the political status quo for those concerned with forging lifestyles and policies compatible with the globe's ecological limits. The costs of caring for an increasingly frail population dominated the concerns of those interested in the problems caused by ageing bodies. Elsewhere, the body was a marker of personal identity for those interested in its rise and status in consumer society, a means of monitoring and controlling the movements of 'dangerous others' for those in charge of national security, and a technologically enhanced and surgically reformed mode of enhancing human capacities for those interested in the weakening boundaries between flesh and machines. In each of these cases, the body assumes a particular meaning, visibility, and value depending upon the perspective from which it is studied.

Uncertainties about what the body is are not entirely new. Worms and snake-like creatures were thought by many in the medieval Christian West to reside within the body, introducing chronic concerns about the stability of the physical organism. Worries about resurrection and the future fates of heaven and hell also encouraged an anxious association between the identities and bodies of believers. Nevertheless, the number of diverse 'stakeholders' in contemporary debates about the body created a situation in which bodily uncertainties proliferated way beyond any single belief system or set of disciplinary assumptions. In this context, the circumstances associated with what many saw as the 'brute facts' associated with humans both *having* bodies (enabling them to act), and *being* bodies (placing certain unavoidable constraints on their activities) were the subject of intense disagreement. Nowhere was this more evident than in debates regarding the physical bases of sex differences.

Chapter 2
Sexed bodies

The idea that our bodies are shaped by social forces and relationships, rather than being ruled and regulated exclusively by natural biological factors, is perhaps most contentious in relation to the subject of sex differences. Indeed, the belief that there exist fundamental and immutable differences in the physiological and neurological make-up of males and females—based significantly on their role in biological reproduction—remains widely held and socially influential. From this perspective, sex differences are evident at birth, amplify during puberty and adolescence, and reach out to influence the personal identities and relationships, leisure preferences and working lives of men and women. They further ensure that the sexes have *fundamentally* dissimilar bodies, tastes, and abilities, excel at different tasks, and are suited to different social roles. As one popular psychological text expressed it, the gulf that separates us is so pronounced that men might as well have originated from Mars and women from Venus.

Space travel is not, however, necessary to the arguments of most who view male and female forms of embodiment as opposites. Rather, it is natural evolutionary processes that are more usually identified as having created the physical, hormonal, and neurological differences between the sexes that determine men's dominance in such areas as physical strength, spatial tasks, and logical reasoning, and women's superiority in multi-tasking, empathy, and communication. Such a chasm, according to those sociobiologists who were influential exponents of this argument from the 1970s, makes it inevitable that the 'facts' of biologically sexed bodies are bound to constrain and direct the organization of society.

Despite the continued popularity of this view, the subject of sexed bodies actually provides us with an excellent means of exploring how social relationships and cultural meanings have, over the centuries, influenced the capacities and destinies of those defined as 'men' and 'women'. The various ways in which 'sex' and 'gender' have been defined and interpreted, indeed, renders problematic the idea that there exist, and have always existed, just two forms of embodiment (male and female). Evaluating the salience of social and cultural factors to this process is, moreover, extremely important: the suggestion that embodied 'sex differences' are natural and unalterable has been used historically to assign fixed identities to men and women, identities that condemn them to limited and unequal roles.

The effect of such stereotypical views of the body can be illustrated by referring to the position of middle-class women in Victorian Britain. Forbidden from entering higher education and dissuaded from participating in vigorous sports, these exclusions were justified by the belief that physical or mental overexertion would damage their reproductive organs and harm the future fitness of an imperial race. Dominated by the natural cycles of menstruation, pregnancy, and childbirth (biological facts that fitted them for a limited role in the home), these women were precluded from those activities and institutions that provided their male counterparts with benefits in terms of their health, social standing, and economic prospects. Anatomy determined destiny.

Male and female bodies in history

Despite the stereotypes that dominated the Victorian era, what is most striking about adopting a long-term historical perspective on sexed bodies is the extent to which leading views on the subject have varied over time. From classical antiquity until the end of the 17th century, indeed, prevailing beliefs about sex differences did *not* involve interpreting bodily contrasts between women and men as opposites, as immutable, or as natural generators of social divisions. Instead, male/female bodies were understood on the basis of what the historian Thomas Laqueur refers to as a 'one sex/one flesh' model. This model was founded on the belief that the bodies of men and women were essentially *similar*, despite possessing limited differences.

During the 2nd century AD, for example, the Greek physician Galen argued that male and female bodies were homologous; an argument reflected in illustrations of the reproductive organs wherein the vagina was depicted as an interior penis and the ovaries interior testes. It was only the excess heat of the male that turned his organs outwards, with the coolness of women's bodies maintaining their inward structure. Such views did not imply that women were judged to be the social, moral, or physical equals to men: for Galen, the extra heat of men made them physically superior to women. Nevertheless, bodily organs themselves were insufficiently stable or different to be a natural cause of individual or social inequalities.

This one sex/one flesh model may now seem bizarre, but it remained the dominant way of thinking about men's and women's bodies for centuries. During the Renaissance, for example, while babies were assigned the status of physical maleness or femaleness according to whether a penis was present or absent, sexed identities were not considered unalterable. The presence or absence of a penis was recognized only as a diagnostic sign of a more complex sexed identity. Just as important to being considered a man or a woman were variables such as whether or not one was considered active or passive, hot or cold, and socially assessed as a 'complete' or 'incomplete' human. In addition, it was recognized that these bodily qualities could change over time, changes that could result in an individual's identity altering from male to female or vice versa. As late as the 16th century, indeed, it was still possible to find anatomists who argued that women could suddenly turn into men if their internal sexual organs were pushed outwards.

These historical examples show us how contemporary views of sexed bodies as opposites have not always dominated people's thinking about the subject. Nevertheless, the one sex/one flesh approach that endured for so long was challenged, and eventually replaced, during the 18th century. Science began to 'flesh out' and make more stable the categories of 'male' and 'female', judging that these referred to natural biological bodily oppositions. From being a malleable indicator of personal identity and social difference, sexed bodies came to be viewed as one of the most important *foundations* for social distinctions, identities, and divisions (determining one's social status as a man or woman, and also fixing one's sexual orientation towards the 'opposite' sex). The idea of the 'inferior female body' as a living organism—but also as a corpse and a skeleton—became of particular interest. It was now generally accepted that women's chaotic and unstable bodies dominated and threatened the rational potential of their easily disturbed fragile minds.

In stark contrast to the 17th-century philosophers Hobbes and Locke, who suggested that there was nothing inevitable about male-dominated social orders, the biological sciences of the 18th and 19th centuries insisted that the limitations of women's bodies ensured their social subordination. Women's natural sensitivities made them fit for producing children, and provided a basis on which they could create a household, but stopped them from assuming significant public roles. Resonating in certain respects with Christian views of original sin, it was thought that childbirth fated women not only to physical pain but also to a highly restricted social existence.

Why did this 'naturalistic' reinterpretation of sexed bodies prove so popular and influential? There is no easy answer to this question given that the previous one sex/one flesh model existed within societies characterized by great social inequalities between men and women. If these earlier societies could do without arguments regarding the physical basis of social inequalities, why did these justifications emerge as important during this later period?

Historians of the body have provided us with one possible answer to this question by interpreting the shift between these views as an ideological solution to one of the key dilemmas arising from Enlightenment thought. The one sex/one flesh model inherited by the Enlightenment created the problem of justifying the continued domination of men over women in a context where progressive philosophical thought was founded on a commitment to equal rights. If the bodily constitution and capacities of males and females were *essentially* similar, despite variations, there was no Enlightenment justification for denying women the rights accorded to men. Perhaps unsurprisingly, this troubled those (men) who stood to lose most by any such possible reform to the status quo. Yet if sexed bodies were not actually malleable biological phenomena, but fixed and unequal organic structures that exerted an unalterable effect on the destinies of men and women, inequalities could be explained on the basis of *natural conditions* over which society was *powerless*. Such naturalistic views of embodiment reinforced men's position in society, while damaging the status of women.

Scientists contributed much to this new argument that sexual anatomy determined social and cultural destiny, often drawing selectively and partially on Darwin's theory of evolution. While women were campaigning for access to higher education during the 19th century, for example, craniometrists (who measured and analysed skulls as indicators of human capacities) ridiculed such aspirations. One of the most influential craniometrists was Gustave Le Bon (a founder of psychology) who argued that the relatively small size of women's heads precluded them from developing mature brains: women were an 'inferior' form of evolutionary development that could never hope to benefit from the educational and other opportunities available to men.

Even when women were granted access to schooling, cautionary voices remained. In America, the President of the Oregon State Medical Society warned in 1905 that intellectual activity could result in mental and physical disease. Relatedly, governmental authorities in England suggested in a 1923 Board of Education report that the 'bodily disturbances' to which girls were subject damaged their mental capacities and impeded their examination performance. There was, furthermore, no let up in the tendency for what passed as science to identify supposedly 'natural' bodily generators of male and female social inequalities.

The most notorious recent example of such accounts is sociobiology. Developed at Harvard University during the 1970s, and based on a reactionary version of evolutionary theory, it became popular as a counterpart to the economically liberal and socially conservative policies pursued by right-wing governments in the USA and UK. Sociobiology held that sexual inequalities were inevitable, constituting the natural and irreversible outcome of genetic differences. Natural selection meant that women evolved to excel in social roles associated with nurturing and caring, while men were designed to compete and dominate. Furthermore, argued sociobiologists, nothing can be gained by attempting to socially engineer society in pursuit of sexual equality: individuals, as Richard Dawkins once famously expressed it, are ultimately no more than 'survival machines' for the real genetic motor force of history and society.

From sexed bodies to gendered bodies

Focusing on how sexed bodies have been classified historically provides useful insights into cultural perceptions of men and women, but it tells us little about how individuals experienced their bodies or could at times use them to mitigate the effects of social stereotypes. In the medieval era, for example, there are records of women gaining respect and recognition within and outside the Christian Church by 'accessing the divine' through dramatic bodily experiences of ecstatic visions, mystical lactations, stigmata, and prodigious acts of self-denial.

Different challenges confronted future generations of women, however, especially once science expounded the view that their bodies were the fragile opposites of men's. The durability and consequences of this 'biology of opposition' were so damaging to women's personal and social opportunities, indeed, that 20th-century feminists engaged in concerted attempts to undermine its credibility. They did this by making a crucial distinction between what was biologically 'given' in the female body and that which was culturally 'added' to it (i.e. those views and prejudices about women that could not be justified by reference to their biological constitution).

These feminist arguments steered a 'middle path' between previous perspectives. On the one hand, by acknowledging that women and men were indeed different as a consequence of their reproductive capacity as well as their primary and secondary sexual characteristics, they recognized the importance of biology to the constitution of sex. In so doing, they shared at least a limited amount of ground with those scientific views of bodily differences that became popular from the 18th century. On the other hand, they sought to restore some of the malleability accorded to male and female bodies evident in the one sex/one flesh model by emphasizing the importance of cultural views of gender. Thus, feminists such as Kate Millett, Betty Frieden, Anne Oakley, and Germaine Greer all emphasized, in distinctive ways, that there was nothing natural about women's bodies that justified either restricting their roles within the family or the sex discrimination they faced in labour market opportunities, rates of pay, or the law. These inequalities, they insisted, had their root in prejudicial cultural views of what it was to be female that could not and should not be justified by reference to notions of the biological body.

Nowhere was this balancing act between recognizing the significance of biology and highlighting the discriminatory effects of culture more apparent than in the writings of Simone de Beauvoir—the most famous feminist of the 20th century. Simone de Beauvoir's 1949 book *The Second Sex* suggested that girls were subjected to 'apprenticeships' into femininity that not only built upon but also *distorted* biological differences between the sexes. These apprenticeships started early: while boys were encouraged to participate in contact sports that taught them self-confidence and how to utilize their bodies for their own ends, girls were directed to passive pursuits that left them vulnerable to being objectified and dominated. For de Beauvoir, it is these *socially organized* activities that limited female bodies, activities that physically socialized women into the restricted roles of wife and mother. As she expressed it: 'One is not born but rather becomes a woman.'

Despite the significance attributed to culture, however, there remains in de Beauvoir's work a sense that female bodies are 'troubled' by their biological functioning. During menstruation, pregnancy, and breastfeeding she suggests that the female body becomes a source of alienation that exposes women to natural forces beyond their control, and can make them feel as if they are 'life's passing instrument'. Nevertheless by focusing on how the capacities of women's bodies are socially mediated, de Beauvoir helped undermine the argument that women's physicality is always naturally and unalterably inferior to that of men.

This distinction between biological sex and cultural gender—with the former being rooted in relatively intractable biological processes and the latter being socially variable—became a cornerstone of feminist analyses. It was on the basis of this distinction, indeed, that subsequent writers opposed to sex discrimination highlighted how cultural stereotypes damaged women's interests by informing popular views of the relationship between women's physical and mental capacities and their social roles. Perceptions of female bodies as weak, unreliable, and lascivious, for example, have been justified historically by such factors as the legal treatment of women's bodies as men's property, and their objectification within pornography.

Western feminists also used this sex–gender distinction when exploring other cultures in order to demonstrate that female bodies did not always condemn women to conventional roles. Anthropological research among the Neur people of East Africa, for example, was used to demonstrate how women could in certain cultures assume the social roles of men. Similarly, female shamans of the Chukchi of north-east Siberia—a culture in which there exist multiple gender identities—occasionally assume male identities and then themselves take wives.

Socially gendered bodies

The Australian sociologist of gender (herself a transsexual woman) Raewyn Connell has taken a closer look at how cultural stereotypes not only alter how women and men are treated, but also shape their physical and neurological make-up. In so doing, Connell has provided us with a valuable general account of the three 'stages' or sets of conditions integral to the development of 'gendered bodies' (bodies that are physically shaped through cultural views and practices). This analysis builds upon the insights of Simone de Beauvoir.

The first set of conditions that must exist for gendered bodies to be created involves the existence of stereotypical views and actions that ignore the similarities, yet highlight and exaggerate the differences, between male and female bodies. This is most obvious in the case of young children who have gendered identities imposed on them—in terms of how they are dressed and talked to, how much roughhouse play they are involved in, and what toys they are given—long before their bodies are capable of engaging in significant actions based on sex differences such as biological reproduction.

The second stage begins when such stereotypical views and actions initiate actual *changes* in the physical development of young people. Boys tend to be encouraged to build up their bodies and impose themselves on their surroundings. Girls, in contrast, tend to be directed towards aesthetic concerns focused on appearance and dieting (a regime that aims to *reduce* the space occupied by the body). These types of activities have very different effects on muscle development and strength. They even have the potential, through their impact on hormone production, to affect bone strength and skeletal development.

The third stage in the construction of gendered bodies occurs when these physical changes are interpreted as *confirming the stereotypes* that helped initiate their development. Women can actually become weaker and less capable at certain physical tasks than men as a result of socially and culturally stimulated changes to their bodies following, for example, years spent dieting and obsessing about appearance. Gendered stereotypes, according to Connell's analysis, are thus rarely harmless: instead, they inform practices that can create real bodily and social inequalities between men and women. This is illustrated in a quotation from the British sports studies scholar Jennifer Hargreaves that reveals how prejudices about frail middle-class women in Victorian England were reinforced by physical inactivity, by corsets so tight that they restricted breathing, and by other aspects of their lifestyle that placed serious restrictions on their bodily abilities:

> Middle class women fulfilled their own stereotype of the 'delicate' females... Women 'were' manifestly physically and biologically inferior because they actually 'did' swoon, 'were' unable to eat,

> suffered continual maladies, and consistently expressed passivity and submissiveness in various forms. The acceptance by women of their 'incapacitation' gave both a humane and moral weighting to the established so-called 'facts'.

The mutually reinforcing effects of cultural stereotypes and physical development are not confined to women or to the Victorian era. Contemporarily, the side-effects of steroid abuse among teenage boys and young men seeking to develop powerful and muscular bodies has become a growing concern, while the prevalence of slim female bodies in adverts and magazines has been associated in recent decades with a growth in eating disorders among increasingly young girls.

The social cultivation of contrasting bodies and physical capacities can also encourage women and men to enter different jobs characterized by very different opportunities and rewards. It is perhaps no accident that while the roles of flight attendant and nurse draw on and develop stereotypically feminine qualities of caring, for example, the requirements of debt collecting or military duties resonate with stereotypically masculine values of intimidation and aggressive physical prowess. Such examples—explored in detail by Hochschild's classic 1983 book *The Managed Heart: Commercialization of Human Feeling*, and in a number of subsequent research projects—demonstrate the very different ways in which men's and women's bodies can be valued in the labour market.

If social prejudices and practices can shape physical development, amplifying or creating differences between the sexes, recent developments in neuroscience suggest that the brain is also likely to be affected by these processes. Speculation about the existence and function of 'mirror neurons', for example, raises the possibility that the gendered changes explored by Connell can reshape the neurological 'wiring' and responses of the brain. Scientists have suggested that mirror neurons help account for our capacity to empathize with others (e.g. almost feeling the pain of another) but it is also possible that this mirroring is *selective* (with neurological 'bridging' between people being stronger among those with whom they have been brought up to identify). It is boys socialized to be physically dominant, for example, who most frequently get 'geared up' to emulate the muscular and aggressive athletes they see on television. In contrast, it is more usually girls who find themselves drawn to lose weight by the airbrushed models that surround them on billboards, in magazines, and on television.

Back to the future?

Exploring the interactions between sexed bodies and gendered stereotyped views and practices has allowed social scientists to move beyond the inadequacies of earlier, scientific theories of men and women as biological opposites. However, science has itself moved on. Indeed, while scientists were responsible during the 18th century for consolidating the idea that the bodies of men and women were biological opposites, advances in endocrinology (the study of hormones) promoted a significantly different view of the relationship between the body and sex from the early decades of the 20th century.

Endocrinologists suggested that the chemicals responsible for sex differences meant that there was no such thing as *the* male or *the* female body: it was more accurate to speak of a *continuum* of sexed bodies. For example, oestrogen and progesterone are often referred to as 'female' hormones, and testosterone as 'male', but women release testosterone from the adrenal gland and males release oestrogen from the testes. What is more, socialization, occupational demands, sporting activities, medication and performance enhancing drugs, and ageing can affect the levels of these hormones within individual bodies. Thus, it is quite possible for a 70-year-old man to have higher oestrogen levels than a younger woman. Bodies here appear malleable, even if not as much as assumed within the one sex/one flesh model of sexual difference.

If endocrinologists provided accounts of sexed bodies that challenged assumptions of biological opposition, developments in genetics also rejected the simplistic sociobiological view that the 'building blocks' of life determine personal identities and social structures. What has been referred to as the 'new' biology, associated with such initiatives as the Human Genome Project, insists that the genetic foundations of human life are far more complex than previously understood. Genes, the hereditary material inside the nucleus of each cell, determine certain physical traits such as eye colour, but more complex characteristics such as gendered identities of masculinity and femininity cannot be reduced to any 'logic' of genetic influence. The significance of genes and groups of genes, indeed, is *codetermined* by interactions within the human organism as well as by interactions that people have with the social and material environment in which they live.

It is not just history and the social sciences, then, but also biology and the life sciences that have developed dynamic and fluid approaches to the subject of sexed bodies. Taking into account the importance of biological and social phenomena previously outside of their main concerns has helped these disciplines construct increasingly sophisticated accounts of how sex is embodied, but this has done little to forge any consensus about what constitutes male or female bodily forms. In the context of increased reflection upon the subject, it has become difficult to ascertain precisely what the sexed body means. Against this background, and in seeking to interrogate further the 'substance' of sex and the limits of its malleability, the contemporary feminist Judith Butler has formulated one of the most radical approaches to this issue.

Beyond sexed and gendered bodies?

Influenced by Esther Newton's (1979) study of American female impersonators, *Mother Camp*, Judith Butler argued that the sexed body attains its social significance and appearance of substance as a result of the *performances* people engage in. These

3. Performing femininity.

performances—that for women utilize props including lipstick, high heels, and jewellery—stylize the body in a manner that approximates to social expectations of what is feminine or masculine (see Figure 3).

According to Butler, however, it is not just cultural notions of gender that are supported and reproduced by such performances, but also the very idea that there even exist such things as 'men's' and 'women's' bodies. When repeated often enough—which they are on a daily basis for women and men who have been 'schooled', socialized, and 'hailed' to speak, look, and act in a manner appropriate to their ascribed sex—these performances manage the physical material they draw upon in a manner that suggests there exists something essential and unalterable about the female or male body.

Butler's focus on structured, gendered performances (phenomena she refers to through the term 'performativity') can help us highlight how the relationship between the body, sex, and gender is fluid even when it appears to be fixed. Butler herself tends to view these performances as prescribed by intractable heterosexual norms governing acceptable presentations of femininity and masculinity, that marginalize people's capacities to challenge them and act differently. Nevertheless, the very idea of performance can be associated with a sense of malleability and it is worth illustrating this point with reference to phenomena of transgenderism. Having experienced a profound dissonance between their bodies and sense of gendered identity, transgender individuals deliberate upon and may decide to change the former in order to match the latter (see Figure 4).

One example of the possible variety of gendered performances involves the case of the world's first transgender international footballer. Born biologically male but having cultivated the appearance of a feminine woman, Jaiyah Saelua is a member of Samoa's *fa'afafine* (a third gender commonplace in Polynesian culture) and in 2011 played for the American Samoa football team in a World Cup qualifying game. Saelua's case illustrates the variety of ways in which a performance can assemble together the body, appearance, and activities in a manner that can provide coherence to an individual's specific gendered sense of self.

Butler's suggestion that sexed bodies can somehow be brought into being by performances is useful in highlighting further one of

4. Transgendered individuals illustrate some of the varied ways in which the embodiment of identity can be configured.

the key themes of this book—the difficulty of specifying what is natural or even 'real' about the body in the contemporary era. However, its overwhelming emphasis on performativity is not without problems. In particular, without attributing greater significance to the biological and physical dimensions of bodies as important dimensions of human beings *in their own right* it becomes difficult to understand the damage done by practices that harm the body. The particular popularity of foot-binding in China during the 10th–13th-century Song dynasty, for example, or modern-day female genital mutilation in sub-Saharan and north-eastern Africa (and elsewhere) are not simply ways of arranging the body to engage in certain performances rather than others. Instead, such customs directly damage the capabilities and potential of women's physical selves.

Despite potential difficulties in Butler's approach, however, her interest in the importance of performances to the 'bringing into being' of sexed bodies remains useful when seen alongside the tendency for women and men to work on their bodies as projects; as raw material to be engineered in relation to particular views of femininity and masculinity. We do not have to 'do away' with the material body to acknowledge the importance of performance: irrespective of what sexed bodies are, they become visible and significant as people manage and mould them in their daily lives, giving off impressions about who and what they are that can confirm or undermine gendered stereotypes.

Gendered bodies and value

The idea that the constitution and capacities of the body are socially shaped, rather than being set by evolutionarily determined natural processes, is perhaps most controversial in the case of sex differences. As made clear by second-wave feminists of the 1960s and 1970s, the idea that women's bodies are naturally fitted for certain limited roles has been defended and deployed regularly by those keen to extract value from women's subordinate position in the household and in waged work. Right-wing governments in both the USA and the UK in the 1970s, for example, developed big-business friendly, 'free-market' policies dependent in part on women's dual role as unpaid housewives and paid workers. They justified these on the basis that their proposals reflected what was natural in terms of biology as well as economics. More generally, men have long profited from women's unpaid labour in the home and women's unequal access to resources ranging from food to jobs. This historical context would seem to suggest that interpretations of sexed bodies are likely to remain a highly contentious issue.

One issue that has appeared repeatedly in critical analyses of the gendering of women's bodies, including the contrasting performances in which they are expected to engage, concerns the schooling, apprenticeship, or more broadly the embodied education through which girls and boys are expected to progress. The suggestion from writers including Simone de Beauvoir, Raewyn Connell, and Judith Butler has been that there are multiple processes of learning and teaching involved in producing bodies possessed of capacities and dispositions that appear 'naturally' female or male. The matter of education, indeed, raises broader issues still about how profit or value is extracted from or added to embodied subjects, and it is to this issue that we turn next.

Chapter 3
Educating bodies

The bodily dimensions of education have long been obscured by those who identify schools as concerned primarily with the mind and with abstract knowledge. This tendency is to be found in the views of both liberal politicians and policy-makers, who equate education with intellectual development and social mobility, and their conservative counterparts, who invest institutions of learning with the responsibility of inculcating into the mindset of pupils socially approved values conducive to future employment.

Such conceptions of education are, however, misleading. Educational institutions do not engage with the mind as a divisible entity—there are no disembodied containers of thought floating around classrooms, workshops, and lecture theatres. They do seek to structure and direct people's *embodied capacities* for experiencing, reflecting on, and engaging with the social, physical, and symbolic environments in which they live. The results of these educational processes can, furthermore, enhance or constrain the ability of people to add value to their own lives as well as to those of others. Feminist analyses of those gendered apprenticeships to which girls and boys are subjected in their passage to adulthood, for example, highlight the restrictive effects of certain educational 'regimes'.

One general way of recognizing that education is a thoroughly embodied process is to insist that there exist *pedagogies of the body*; pedagogies concerned with the inextricable relationship between practical physical action and thought. Body pedagogies, or pedagogics as I have referred to them, consist of ordered sets of practices designed to cultivate particular techniques, skills, and sensory orientations to the environment; phenomena themselves associated with specific types of knowledge and beliefs. For the sociologist and anthropologist Marcel Mauss, indeed, it is the social methods by which bodies are educated that turn human organisms into particular types of persons able and willing to function practically and intellectually in the cultures to which they belong.

Most people are exposed to institutionally sanctioned forms of body pedagogics early in life. Since the post-war development of the welfare state in the UK, for example, schools have complemented families in being *the* prime sites in which occurs the monitoring, caring for, and maturation of large numbers of children. This education explicitly targets embodiment as a whole: from the earliest stage of primary education pupils receive each year hundreds of physical corrections to their movements and actions by teachers concerned not only to stimulate their capacity to think but also to instill in them particular ways of managing their bodies, interacting with others, and experiencing their surroundings. Teacher attempts to get children to dress themselves, to visit the toilet before accidents occur, to respect the personal space of others, and to participate in daily rituals such as morning prayers, demonstrate the fundamental importance for schooling of educating the *moving* and *managed* body, and not just some abstract conception of the 'thinking mind' (see Figure 5).

Bodily pedagogics are not confined to countries that possess modern welfare states. They have existed at every stage of human history. Spartan males in Ancient Greece, for example, were enrolled from the age of seven into the *agoge*; a harsh and prolonged regime of education and training directed towards producing skilled and resilient soldiers. More recently, the significance of

5. Body pedagogics involve the disciplining and cultivation of children's abilities inside and outside the classroom.

'muscular Christianity' in Victorian Britain (also evident during the latter half of the 19th century in America and Australia) focused on developing through team sports and disciplined physical education a ruling class fit to govern its colonies.

The education of bodies in and beyond schools has a long and varied history, but its diverse manifestations have in common the attempt to develop and steer in a particular direction the energies, capacities, dispositions, and deliberations of those subject to them. Seeking to survive and prosper in an increasingly competitive global environment, contemporary nation-states frequently continue to place education at the centre of their efforts to ensure that future generations of workers are as physically and mentally productive as possible. Echoing the earlier approach of a professor of physiology from Cornell University who prized body management as an essential expression of patriotism (by promoting dieting after estimating that New Yorkers alone carried ten million pounds of excess fat that would be better used as rations for soldiers during World War I), politicians from every continent have embraced the idea that the education of bodies can be both a vehicle for control and a potential source of value for society.

In the context of these developments the educational theorist Basil Bernstein has suggested that modern cultures are becoming 'totally pedagogized': it is not just national educational systems but also a large number of trans-national and more local social, cultural, and religious groups who are promoting as 'normal' and desirable their own versions of what form education should take. If bodily pedagogics have become a matter of such importance in the contemporary era, however, we need to examine in more detail what is involved in the cultivation of people's embodied being.

Techniques of the body

Writing in the first half of the 20th century, Marcel Mauss addressed in detail the mechanisms involved in body pedagogics via his suggestion that societies exist as places in which people can interact effectively with each other only by successfully transmitting to each new generation shared '*techniques of the body*'. These techniques consist of particular ways of utilizing the body when it is engaged in the full range of activities over which it is possible for an individual to exert control—activities ranging from the instinctive but apparently socially insignificant, such as breathing, walking, and squatting, to the most aggressive, involving combat. Integral to acquiring any and all such techniques for Mauss is a *biological process* involving the posture, musculature, and movements of the body, a *psychological process* involving receptivity, effort, and reflection on the part of the individual, and a *social process* of apprenticeship and imitation characteristic of efficient learning and teaching. Mauss's analysis enables us to see how it is that education provides a means whereby the various dimensions of human embodiment can be utilized to attach individuals to cultures.

Specific techniques of the body usually survive and prosper most effectively when they assist the goals of individuals as well as the objectives of the collectivity. While certain body techniques may be common to a society or culture, however, others serve to differentiate between people. Sometimes this occurs when the dominant social classes are able to classify certain techniques as low status and to be avoided wherever possible. This was evident in 18th-century Europe when the ruling classes treated walking as a mode of transport fit only for the poor, the criminal, the young, and the ignorant. In other cases—as in body techniques associated with masculine physical violence, on the one hand, and feminine caring and domestic labour, on the other—their unequal distribution can help men control, dominate, and extract value from women. There is also a class component to techniques of violence. During the feudal and early modern era, for example, propertied men were usually more of a threat than peasants since they possessed the weapons, skills, and also horses that others lacked.

Techniques of the body can also serve to differentiate between people because of the sheer degree of expertise and prolonged training with which they are associated. The body may be our 'first and most natural instrument', but the complex processes involved in high-level skill acquisition mean that these abilities are often available only to those possessed of the means to undergo prolonged apprenticeships (a variable also related to the opportunity structures of society). Becoming a professional musician or athlete, for example, necessitates that muscular responses, senses, and sensitivities are all adjusted to the relevant task, and to each other, before progress can be achieved. Even actions such as manipulating a violin's strings, or grasping and moving a cricket ball across the palm and between the fingers prior to delivering it towards the stumps, are subject to a prolonged learning process as a result of the complexity of the hand and the subtly different ways in which holding, pinching, cupping, and cradling objects can occur.

Reaching the stage whereby such adjustments can be made not only in executing a specific task, but also across a range of cognate tasks central to a trade, sport, profession, or vocation can, indeed, require an apprenticeship lasting years. It is notable that various researchers have suggested that it takes around 10,000 hours of sustained and purposeful practice to become a skilled craftsperson, for example, or proficient in any higher level skill.

Philosophers have accounted for the difficulties associated with acquiring these skills by detailing how this process always requires overcoming gaps—gaps between the existing abilities of the individual intent on undertaking a challenging task, and the objects or environments s/he is attempting to manipulate. Useful empirically informed analyses of this bridging of gaps between early attempts and later successful execution of a task have been provided in the cases of sailing, physical education, and craftwork by Leif Östman and his colleagues at the University of Uppsala. A more mundane example can be given in the case of home improvement. When drilling into an uneven wall, hammering a nail in an enclosed space, or attempting to saw at a precise angle through a plank of wood, it is necessary to engage in adjustments involving a carefully balanced alignment of senses and actions with the properties of the relevant raw materials. For the amateur such as myself, the slightest irregularity in a wall can and indeed does cause a problem when drilling is done mechanically without a sense of the minute variations required to maintain the constancy and direction of the hole, and to avoid the drill bit moving away from the true line, skidding across the wall. For the expert, in contrast, adjustments of posture and pressure occur automatically through a sensitive responsiveness to the actions being undertaken and the materials being worked with.

Achieving expertise in an area of practical accomplishment can have a profound impact on embodied experience. Successfully mastering skills involves a 'pouring of the bodily self' into the task at hand; a process characterized by a sense of harmony between the individual subject and the external objects. From struggling with a task that appears alien, awkward, and disconnected from the participant, a sense of immersion and ease, or what psychologists refer to as 'flow', can emerge from the successful execution of a skilled technique. This is evident in the sociologist David Sudnow's account of learning to play jazz piano: tortured attempts to create music gradually gave way to a capacity to improvise effectively. Another consequence of successfully mastering such skills is that there emerges a distance between the previous and present bodily self, and a new distinction between those with whom one now shares this skill, and those bereft of such abilities; a distinction that has important consequences when it comes to social stratification and the development of social identities.

These processes and experiences, together with a sense of the social distinctions and divisions to which they may give rise, can be explored further with the assistance of examples taken from occupational training (involving the education of vision), from religion (involving the charismatic Christian 'Alpha Course'), and from sport (involving the apprenticeship needed to become a boxer). Taken together, they illustrate how it is the educated body that provides us with what the philosopher Merleau-Ponty refers to as 'our vehicle of being in and our medium for having a world'. Given the various ways in which these body pedagogics develop the practical, sensory, and intellectual capacities of the embodied subject, however, they do not necessarily help us to provide a single answer to what the body is. Instead, focusing on the diverse ways in which bodily subjects can be pedagogically 'enhanced' by social relations and training, they direct our attention to the body as a resource, to how the body is valued, and to questions of what the body can do.

The occupational training of sight

Sight was until recently often neglected in discussions of the body pedagogics involved in occupational education and training.

During the past decade, however, a number of studies have examined how the acquisition of specific visual techniques constitute an integral part of particular professions. These are well represented by fascinating yet contrasting sets of analyses into how animals are seen and judged, on the one hand, and how experts learn to see and interpret medical images, on the other.

The social and behavioural scientist Cristina Grasseni has undertaken research into how cattle breeders and judges learn to rank and assess excellence among purpose bred cows. Two things stand out from her account. The first concerns how visual judgements shaped by professional standards of size and proportion are learnt only after considerable education and training. The length of this 'induction' is such that the children of breeders will begin learning their craft by playing with plastic toy cows that replicate aesthetically preferred standards. Learning what to look for in a cow—such as the size, texture, and proportion of key body parts and the relative importance attributed to each of the variables that constitute excellence—takes practice and guidance.

The second feature of her account that is especially notable concerns how these visual standards and practices have over the decades—through breeding practices that select and partner cattle on the basis of these aesthetic standards—served to *transform the appearance of the animals*. Far from being passive, vision is in this case tied to bodily practices that *re-make* part of the animal world. In this case, reflections on and assessments of the quality of animals' bodies inform practices that affect the future development of those bodies.

The other study I want to refer to here that explored how animals are seen resulted from the anthropologist Rane Willerslev's fieldwork into the Yukaghirs, a small Siberian indigenous group. Willerslev details how these hunters must learn to take on the 'viewpoint of elks' in order to track them effectively. Instead of having to acquire and see through the lens of breeders' standards, the Yukaghirs demonstrate how learning to see effectively can be dependent upon achieving a degree of *cross-species* communication. The body pedagogics of tracking elks requires the hunter to see while moving with great care through challenging terrain, and also to be aware of and responsive to *being seen by* the prey.

In maintaining what Willerslev refers to as this 'double awareness', the hunter must seek to 'adopt the perspective' of the elk by appearing unthreatening—an impression which involves mimicking passive postures while maintaining the intent and gaining the proximity that allows for the prey to be killed. Far from being a passive sense, sight is again linked to an education of the body that shapes how individuals (in this case hunters) orientate themselves towards and act within the world. Learnt through guidance from elders, and the unforgiving feedback from elks that run from clumsy attempts to track them, the eventual achievement of proficiency also facilitates a shift in social status. Being able to navigate the tensions between seeing and being seen by the elk opens up a new world of possibilities for the hunter; it enlarges their potential for acting on their environment, while also earning respect from others.

Very different examples of educated vision are provided by studies into the interpretation of computer tomographic (CAT) scans. These scans produce medical images of any section of the head and body in an attempt to detect anomalies associated with such factors as tumours, injuries, or abnormal blood vessels. These images are by no means always clear indications of what is happening within the body, though, and can appear as a collection of dark blobs, and patches of light and shadows that are meaningless to a novice.

The prolonged education needed to decode these images requires a shaping of seeing that is guided by those rules, procedures, and intuitive interpretive jumps acceptable to the medical community. Interpretation is needed in order for meanings to be attributed to an image that would otherwise appear incomprehensible, for example, while a process of translation is needed to judge their precise physiological or neurological location and significance.

Despite the differences between looking at animals and looking at medical images, these examples complement each other in demonstrating how vision can be trained to see new things, to yield new kinds of valuable information, as part of a wider professional education in body techniques. These forms of education may be directed towards very different outcomes, but they harness 'ways of seeing' to particular types of acting upon and transforming the environment (be this transformation concerned with the breeding of cattle, the hunting of elk, or the health of patients).

An apprenticeship in boxing

The processes involved in a very different form of body pedagogics are the subject of the sociologist Loic Wacquant's 2004 book *Body & Soul: Notebooks of an Apprentice Boxer*. Wacquant did not initially intend to study boxing: he made contact with Woodlawn boxing gym in Chicago when seeking to meet people who would help his study into the black American ghetto. Yet early experiences prompted him to undertake a participant observation analysis of boxing, and thus began a three-year project involving him training three to six times a week alongside amateur and professional boxers.

Wacquant's aim was to chart the processes involved in acquiring the 'pugilistic habitus' (the techniques, habits, ways of seeing, feeling, and responding that define the competent boxer). What is particularly interesting about his account is the gap he describes between his clumsy initial attempts to follow the routines of training, and his later accomplishments after hours of dedication, hard work, and pain began to pay off. These initial efforts are illustrated by his early summary of sparring:

> My lungs are about to explode; I don't have any legs or strength left. I follow him, jabbing in a fog of fatigue, sweat, and excitement. My fists are quickly growing too heavy, my arms numb…I'm losing my energy at lightening speed and my punches aren't snapping any more.…'Timeout!' Finally it's over! I am at the edge of asphyxiation, tetanized with exhaustion, totally drained in 6 min. I feel like I'm going to vomit up my lungs and pass out.

How are the bodies of novices such as Wacquant educated and transformed in order that they can incorporate the techniques and cope with the demands of this sporting art? The formal start of this process begins in the gym; a space and culture that promotes a regimented lifestyle of diet, sleep, and exercise such that boxers often compare it to entering the military. Training itself includes shadowboxing, abdominal exercises, skipping, and other disciplines that gradually instill in the beginner techniques, coordination, and a stamina suited to surviving this regime. Pedagogically, the satisfaction of being told by the trainer that one is doing something correctly complements the pain of being hit when techniques prove ineffective in sparring sessions. Emotions must be controlled in order to minimize punishment in the ring, and to inure one from deciding to quit in the face of adversity, and there is no substitute for the long hours of repetitive actions it takes to educate the body into the orientations required to be a boxer.

In Wacquant's case, improvements came with time, and his account of this apprenticeship (that culminated in a competitive bout) details a period of progress in which sight, movement, resilience, and stamina are closely aligned:

> From session to session my field of vision clears up, expands, and gets reorganised: I managed to shut out external calls on my attention and to better discern the movements of my antagonist, as

> if my visual faculties were growing as my body gets used to sparring. And above all I gradually acquire the specific eye that enables me to guess my opponents attacks by reading the first signs of them in his eyes, the orientation of his shoulders, or the position of his hands and elbows…

To undergo and assimilate the body pedagogics associated with boxing is, of course, a very different experience from the visually led practices discussed in the previous section. Nevertheless, it reinforces the point that education is a thoroughly bodily phenomenon, and that ways of seeing are integrally related to, and not separate from, physical capacities and activities. Wacquant also makes the point that the practices involved in learning to become a boxer can exert a profound effect on thought processes. Boxers learn *from their bodies*, as well as from the bodies of others, that the strict pursuance of diet and physical training enhances the chances of sporting success. They come to know *through* the physically informed culture of the gym that late nights, alcohol, and over or undertraining increase their chances of getting hurt.

The routines, excitements, and *esprit de corps* associated with this sport even begin to shape the dreams of those who persist in honing their skills. Wacquant himself experienced this when confronting the prospect of having to renounce the boxing lifestyle to which he had become accustomed in order to return to the physically sedentary life of academia. Having redirected his actions and embodied thoughts to the lines of activity and aspiration embedded within the body pedagogics of boxing, the prospect of immersing himself in the relatively isolated business of translating thoughts on to paper and writing a thesis seemed depressing.

Becoming religious

The third example of how bodies develop and are rendered distinctive from each other through the acquisition of socially and culturally patterned pedagogics is provided by the religious studies scholar Tony Watling's research into the experiences of those who take the Alpha Course. This is a fifteen-week programme associated with evangelical Christianity and designed to prepare the initiate to receive the Holy Spirit during an intense concluding weekend. As such, it is meant to help individuals enter a direct relationship with God, and is associated by those who designed it with the spiritual condition of purity characteristic of the 1st-century Christian Church.

Alpha programmes grew in popularity during the 1990s and now deal with millions of people per year across a wide range of countries. Individual courses, however, are often intentionally small affairs, with a dozen or so participants tutored by several 'leaders' who are committed Christians, who have 'received' the Holy Spirit, and who guide the education of participants. The course itself aims to inform and impart to its students an intellectual understanding of the major principles of Christianity, but underpinning and reaching beyond this cognitive goal is the aim of educating the bodies of initiates in order that they will receive an intense emotional experience that provides a basis for this understanding and also encourages personal identification with Christianity.

One of the pedagogic techniques used to stimulate this experience witnessed by Watling involves initiates being sat on a stool in the centre of a room while others lay hands on head and shoulders, praying for the Holy Spirit to fill that person's body and mind. The manner in which people experience the Holy Spirit varies—ranging from a sudden sense of possession manifest in fainting, uncontrollable laughing, crying, or speaking in tongues, to a dawning realization of presence during prayer. Nevertheless, it is common for these feelings to be associated with a sense of 'release', 'surrender', and even rebirth, and a new determination to embrace God as mentor and guide.

While Watling's case-study reveals how the Alpha Course possesses its own particular means of educating the bodies of initiates, it supports Marcel Mauss's arguments about the widespread existence of religious body pedagogics. In his comments on Eastern religiosity, Mauss identifies the significance of breathing techniques to both the experiences of cosmic balance and harmony enjoyed by Taoist priests, and the transcendent contact with the divine enjoyed by Yogic mystics. Examples such as these suggest that religious experience is, at least in part, dependent upon the type of education received by bodies.

If it is possible to identify body pedagogics across varied religions, it is also important to note that contrasting techniques promote very different religious experiences, practices, and beliefs. Christianity and Islam are both monotheistic religions, for example, but their respective pedagogics are associated with very different outcomes. Christianity has generally promoted ritual techniques that emphasize *communion* (e.g. through baptism and participation in the Eucharist); promote a sense of *individual transcendence*; and seek to call believers *out of this world* in a process of rebirth that accepts a distinction between this-worldly imperfection and other-worldly perfection. In contrast, Islam has generally emphasized body techniques indicative of *submission* (e.g. through the gymnastics of prostration that accompany *salat*/five-times-daily prayer); that seek to stimulate a sense of *collective identification* with other Muslims (as reinforced by the bodily fasting required during Ramadan and the privations traditionally associated with pilgrimage); and that seek to promote a 'total society' in which divine law (*sharia*) helps ensure that worldly realities correspond to transcendent religious goals.

The 'known' and the 'unknown'

Education is as much to do with bodily doing as it is with cognitive thought, and while the two phenomena are perhaps ultimately inseparable, the former has tended until recently to be neglected in most analyses of teaching and learning. This neglect is important not only because it marginalizes issues related to how we learn to engage with, experience, and alter the environment in which we live, but also as a result of how the development of physical abilities, habits, and techniques informs our reflections on and beliefs about the world.

In contrast to the traditional Western view that our minds possess independence from the activities and relationships in which we are enmeshed—defining us as human because of their capacity to facilitate thought—practical experience and action is not necessarily opposite to, or undermining of, intellectual insight and reflection. Irrespective of whether this involves judging or tracking animals, scrutinizing medical images, communing with God, or any number of other examples, the bodily techniques through which we experience and act on the world can facilitate a deeply engaged type of thinking informed by the nature of the environment in which the individual intervenes.

Developing this point, the pragmatist philosopher John Dewey explains the links between bodily experience and action, on the one hand, and conscious mental thought, on the other, through the distinction between, and connections that link together, *anoetic* and *noetic* knowledge. Anoetic knowledge exists independently of conscious thought: it inheres within our embodied selves outside of conscious attention as an awareness, intuition, or knowledge of how to do something yet to be formulated reflexively. Noetic knowledge, in contrast, is manifest consciously in terms of reason, reflexivity, and intellect. While it is often difficult and sometimes impossible to translate sensory-perceptive experiences into noetic knowledge, they form a layer of being on which thought and reflection are formulated.

Dewey's discussion is important as it suggests that the education of bodies operates at the level of pre-conscious anoetic knowledge as well as conscious noetic thought. Looked at from this perspective, the examples of bodily education examined in this chapter involve the shaping of particular kinds of preconscious awareness, assumptions, and orientations. The differences that exist between the sportsperson who 'feels at home' when engaged in rigorous physical training, the religious mystic for whom stillness is preferable, and the musician who feels most alive when playing cannot be explained solely on the basis of intellectual knowledge or values. Instead, body pedagogics provides an anoetic and pre-conscious basis on which understanding develops.

This approach to educating bodies has important consequences not only for understanding how cultural differences emerge between people, but also for appreciating some of the difficulties that can confront communication and dialogue between groups of people seeking to overcome conflict. If we accept that different forms of body pedagogics create gaps between people—in terms of both their practical and intellectual sense of what is natural, feasible, and desirable—then the idea that talking on its own might prove a means of understanding will often be of limited value. Body pedagogics are inextricably related to particular cultures, and the history and traditions of these cultures, and in the case of religious differences, for example, it should not surprise us if these inform intractable clashes.

Issues concerning which God to pray to, how to pray, what to eat, whether evangelical activity is an option or a duty, and whether religious pluralism or even cartoons of God are acceptable are not just matters of belief. They are also inscribed on how people have learnt to use their bodies and engage with the environment dealt with by religion. This point is developed by the anthropologist Tim Ingold's exploration of different body pedagogics. By encouraging a 'communion of experience' among people, by immersing them in the same types of physical learning, he suggests that those responsible for what we might call the 'corporeal curriculum' of a group, culture, or nation are cultivating an embodied foundation on which attempts at verbal understanding and discussion must build.

There is, of course, no guarantee that any particular individual will acquire successfully the skills and aptitudes associated with a particular education of the body. Just as many pupils reject as irrelevant or are unable to learn the academic knowledge presented to them at school, so too do many people fail to internalize the body pedagogics to which they are exposed. One particularly worrying example of the unintended consequences of body pedagogics is provided by research conducted by the physical education specialist John Evans and his colleagues at Loughborough University into the effects on highly motivated young women of recent education initiatives designed to increase physical fitness, enhance sporting success, and combat the 'obesity crisis'. These initiatives have reinforced trends within consumer culture that seek to deify unfeasibly thin females, and Evans has uncovered a disturbing tendency for these initiatives to contribute towards anorexia and other eating disorders—to subtract from rather than add to the capacities for action of those affected. Such research suggests that while people's identities, and the manner in which these vary both historically and cross-culturally, are integrally related to the forms of embodied education they receive, these body pedagogics can produce significant and unwanted consequences.

The processes involved in educating bodies are ubiquitous throughout society, even if they are most visible within formal educational institutions. They raise important questions about who gets to select and control the embodied techniques, information, and values transmitted to individuals within distinctive cultural settings. The sheer diversity of ways in which embodied individuals and groups can be educated to use their bodies, senses, and minds does little to clarify what we mean by 'the body', but it does direct our attention to equally important matters regarding what the body can do. Body pedagogies not only steer the development of people's capacities in certain directions but can also, depending on the form they take, add value to the capacity of individuals to engage with and change their environment. If contrasting forms of body pedagogics add value to individuals in different ways, so too do they show how what is valued and prized about bodies and their sensory capacities can vary enormously.

Chapter 4
Governing bodies

Theories of governance—of how rule or regulation is accomplished across nations, institutions, and organizations—have long included a focus on the strategies states pursue in seeking to control and develop the bodily capacities of their citizens. In so doing, they raise questions about how states conceptualize the bodies over which they seek to govern, about those aspects of people's embodied being they value, and about the resistance they may confront when attempting to implement their policies and achieve their aims.

In analysing these issues, the sociologist Bryan S. Turner has gone so far as to suggest that the very fact citizens are embodied beings confronts governments with unavoidable challenges. These involve: (1) ensuring viable population levels over time; (2) regulating the movement of bodies into, out of, and within its territories; (3) ensuring that the physical expression of people's desires is compatible with peaceful coexistence; and (4) maintaining a basic consensus about how individuals should present themselves to, and interact with, others. For Turner, these challenges have tended to be addressed through systems of delayed marriage and patriarchy; through surveillance, policing, and record-keeping; through socialization processes that prize self-control; and through enduring customs regarding appearance and behaviour.

Turner provides us with an influential view of the relationship that exists between government and embodiment, and also points our attention to how these links have been explored historically. In Ancient Greece, for example, Plato portrayed the citizens of the Republic as divided into three classes on the basis of their physical and moral constitution. Aristotle developed these ideas of government and bodily inequalities by reasoning that while politics should promote noble physical and mental habits among residents of the polis, the bodily natures of women and slaves precluded them from developing these virtues.

Centuries later, very different conceptions of government as a *social contract* between rulers and ruled continued to be concerned with the essential nature of embodied individuals. Writing against the background of political turmoil and civil war, the English philosopher Thomas Hobbes argued in his 1651 book *The Leviathan* that individuals naturally gravitated towards violence in the 'state of nature'. For Hobbes, this destructive behaviour would invariably result in a war of 'all against all' unless people agreed to cede fundamental rights to the state—rights that handed the power of life and death to the government.

Hobbes and other social contract theorists of the early modern era such as Rousseau were concerned to specify the conditions in which embodied life could be regulated, preserved, and enhanced given what they assumed to be the physical desires and propensities at the heart of human nature. While recent writers on the body and governance objected to the assumption that there existed an essential human nature, they have continued to explore the relationship between state control and behaviour. The 20th-century sociologist Norbert Elias, for example, argued that the growing capacity of states to punish unauthorized acts of violence was associated with the spread of peaceful forms of human interaction and the more considered expression of individual desires.

The rise of biopower

Seeking to comprehend historically and theoretically those changes that have occurred in the relationship between state rule and the body, the French philosopher Michel Foucault focused on the transition from medieval to modern forms of governance. For Foucault, this transition revolved around a change in the nature and 'an explosion' in the number and variety of what he refers to as 'biopolitics': 'techniques for achieving the subjugation of bodies and the control of populations'. The cumulative effect of these shifts replaced the medieval focus on death with an emphasis on exerting control through the positive management of *life*.

The medieval focus on death

In medieval absolutist states—states in which the monarch exercised complete authority—the legal treatment of bodies tended to operate on the basis of a simple distinction between life and death. There were few central attempts to regulate in detail how law-abiding people lived their lives, but threats to the security of the state provoked spectacular displays of monarchical power involving torture and destruction of the offender's body (see Figure 6).

During the late early-modern period, however, major changes took place in the space of a just a few decades that shifted the focus of governance from death to life. Foucault illustrates this in a gruesome opening passage to his book *Discipline and Punish* in which he details the 1757 torture and execution of Damiens (convicted of attempting to assassinate King Louis XV of France). Tied to a scaffold, Damiens had lumps of flesh torn from his body with red-hot pincers before a 'boiling potion' was poured over each wound. Screaming from this torment, his limbs were then wrenched and ultimately separated from his body. Finally, while still alive, his trunk was thrown on to a stake and consumed by

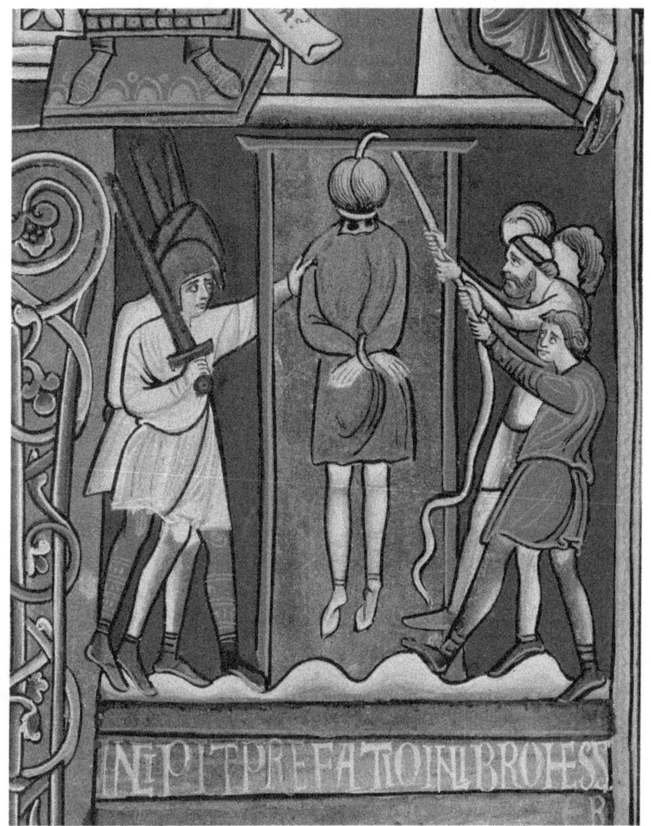

6. Stories of Esther; the hanging of Haman, miniature from the Bible of Souvigny. Latin manuscript 1 folio 284 recto, 12th century.

fire. This execution occurred in plain view of, and was a source of entertainment for, spectators: it was a show to be enjoyed.

If governance in the late early-modern period maintained elements of the medieval focus on torture and execution, however, developments during the late 18th and early 19th century brought about a fundamental shift away from rule based on gross displays of physical violence. Concerned by the 'habitual indifference' promoted by public spectacles such as Damiens' execution, social reformers and politicians became worried about the obstacles such events placed against the maintenance of social order. In place of this violence, there was a gradual move towards incarceration in conditions considered conducive to restoring offenders to society. It was this shift in governance that laid the foundations for the modern biopolitical management of life.

The modern focus on life

The modern approach towards the management of embodiment involved for Foucault a new 'art of government' in which disciplining the 'soul' (or the inner desires, habits, and actions of the individual) became more important than punishing the physical flesh. This art was focused upon improving the quality of the population in terms of valuing its 'human capital' (the behaviour, health, education, skills, and capacities of people insofar as these are relevant to the creation of economic and other forms of profit). It was made possible by the development of new institutions and the knowledge with which they became associated. For example, the proliferation of asylums and hospitals, and innovations in the organization of factories, emerged alongside advances in biology and modern political economy. Innovations in the modern army facilitated the introduction of IQ tests, and ideas of average levels of intelligence. These circumstances enabled governments to classify and seek to manage individuals on the basis of notions of what was 'normal', efficient, and productive—or alternatively 'abnormal', inefficient, and unproductive—for the broader groups to which they belonged.

Against this background, the governmental art of enhancing human capital assumed a variety of forms involving the assessment and management of people, yet two of these were especially prominent. The first involves what we might term the 'productive surveillance' integral to 18th- and 19th-century ideas of

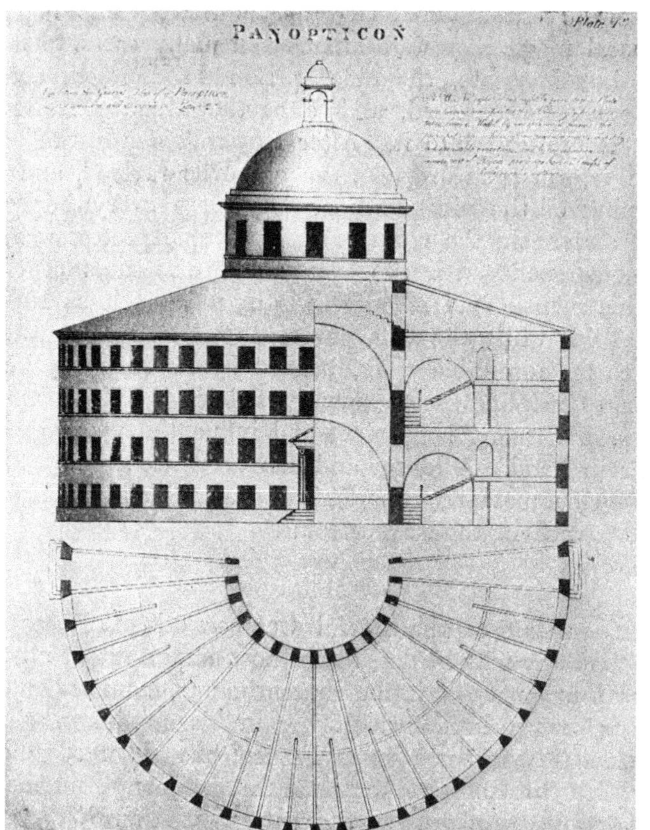

7. Bentham's panopticon.

prison reform and workplace organization. In the case of prison reform, this was exemplified by the English philosopher Jeremy Bentham's design for a 'panopticon' (see Figure 7). This involved a central watch tower from which prisoners could be monitored. With their every action open to scrutiny, this visibility was meant to encourage prisoners to reflect on their behaviour from the perspective of the warden, and to improve their self-control. The medieval concern with torturing and dismembering the criminal's body was here replaced by attempts to achieve rehabilitation; a step that, repeated across the prison population, had the potential to enhance the productive capacity of society as a whole.

Productive surveillance was not confined to prison reform. Indeed, the plan of installing a central surveillance point overlooking a much larger area was first developed in the 1780s by Jeremy Bentham's brother, Samuel, while working in Russia for Prince Potemkin. Here, the panopticon was designed to enable a small number of factory managers to supervise and increase the output of a much larger workforce. Foucault provides us with an example of how these principles were actually implemented when describing how managers at the late 18th-century Oberkampf printing factory in Jouy, France, organized tables on the ground floor into rows. By walking in between them, the supervisor was able to reduce unproductive conversation and interaction between workers, as well as compare their application, speed, and quality of work.

Factory-based productive surveillance developed further during the era of 'scientific management' based on the work of the mechanical engineer F.W. Taylor and originating during the 1890s in the USA. Separating the conception from the execution of work, and timing the fastest way to complete tasks, Taylor suggested that this scientific approach would enable employers to set workload norms and maximize their profitability (see Figure 8). Such ideas of governance built upon increasingly influential models of the body as a motor. The German physician and physicist Hermann von Helmholtz, for example, applied this view to machines and humans, encouraging the latter to be viewed in terms of the former. Once again, the way in which the body was conceptualized and valued is shaped by the social context in which it is cultivated and managed.

The second area in which biopolitical attempts to enhance human capital were particularly prominent was eugenics.

8. Production lines can facilitate the surveillance of workers.

'Eugenics' was a term developed in the early 20th century by Francis Galton, who applied Darwin's theory of evolution to proposals for improving the future quality of the human race. These ideas received ready support as 'national efficiency' movements in Britain and America had already expressed concern about the reproductive and physical fitness of the population. Problems associated with the poor quality of recruits to the British military during the Boer Wars of the late 19th century, for example, led to the establishment of the Committee on the Physical Deterioration of the Race.

Eugenic theories were evident in the development of the international birth control movement. One of its early influential figures, the American maternity nurse Margaret Sanger, argued before World War I that the modern era was threatened by unwanted pregnancies. In a related development, the first clinic in Britain allowed to dispense contraceptive advice and products without legal restriction, established by Marie Stopes in 1921, was justified on the basis that only those able to 'add value' to the race should become parents.

The growing influence of eugenics was evident in the arguments of David Lloyd George, Britain's Prime Minister from 1916 to 1922, who encapsulated the widespread worries about the 'degeneration of the racial stock' by stating that it was not possible to run an A1 Empire with a C3 population. Eugenics even became an academic discipline at many universities, and its supporters in Britain alone included the economist John Maynard Keynes, the zoologist Julian Huxley, and the psychologist and pioneer of IQ testing for school children Cyril Burt. Eugenic policies themselves reached a high point in the years between the two world wars. By this time, the forcible sterilization of patients deemed mentally defective was occurring in a range of countries, including Sweden, Japan, and Canada. In America, by the onset of World War II over 40,000 people had been sterilized involuntarily.

The popularity of eugenics waned after the discovery of the death camps in Nazi Germany; phenomena that demonstrated in extreme form how a concern with the 'health', 'strength', and 'purity' of the population could also involve the systematic management of death. Nevertheless, policies and practices designed to limit population growth and eliminate hereditary diseases and disabilities of all sorts continue to be both popular and controversial, and are associated now with biotechnologies that did not enter even into the hopes and dreams of the eugenicists. In China, for example, there still existed legal prohibition against procreation among those with a number of hereditary conditions at the end of the 20th century.

The two examples of biopolitics examined in this section—productive surveillance in prisons and factories, and eugenic strategies for 'racial improvement'—seek to exert control over distinctive aspects of embodiment. While the focus on human capital is concerned with control and productive behaviour among the *present* population, the focus upon reproductive activities exists in terms of increasing the quality and value of *future* generations. In these cases bodies are viewed quite differently, and are related to very distinctive political goals, yet the Italian political philosopher Giorgio Agamben suggests that they also converge in certain vital ways.

For Agamben, indeed, the very development of biopolitics represents a move away from the political concern with noble action and conceptions of the 'good life' characteristic of the Polis in Ancient Greece. Instead, it constitutes a trend towards a preoccupation with what he refers to as 'bare life', or the simple facts of living common to all groups. It is these basic life processes that increasingly preoccupy states according to Agamben; processes that are relevant insofar as they are seen to relate to notions of productivity, ability, and disability, and that have been condemned by critics as rendering inconsequential the genuinely political realm. The reach and variation of biopolitical policies concerned with bare life has, moreover, increased even further in relation to security and medicine, raising anew questions about what the body is and how it is consequential for the governance of society.

Contemporary governance

In exploring how the character of biopolitics has changed within the 20th century itself, Foucault argues that the rapid development of capitalism was associated initially with relatively 'heavy, ponderous, meticulous and constant' disciplinary regimes in schools, hospitals, barracks, factories, and families. Yet as the 20th century wore on, control was exercised through more sophisticated techniques. With the growth of consumer culture in the second half of the 20th century, for example, Foucault finds evidence of 'a new mode of investment which presents itself no longer in the form of control by repression but that of control by stimulation. "Get undressed—but be slim, good looking, tanned."'

Despite Foucault's emphasis on control through productive stimulation, however, both coercive and 'softer' forms of bodily control and regulation clearly still exist. This is evident when we explore how contemporary forms of surveillance (deployed increasingly for the purpose of national security) have very different effects on contrasting groups of people.

Security and biopolitics

Techniques of systematic surveillance, anticipated by the idea of the panopticon, underwent a massive expansion in America following the Bush administration's announcement of a 'War on Terror' immediately after the terrorist attacks of 9/11. Since then, the Central Intelligence Agency, the National Security Agency, the Department of Homeland Security, the Federal Bureau of Investigation, and a range of other state and quasi-state organizations have received billions of dollars of public funding enabling them to observe, trace, and record the data trails left by people both inside and outside the USA.

These projects have been complemented by 'fusion centres' that pool together and share the vast amount of information gathered by these agencies: the extent of these activities is such that US spying activities have been collecting and storing over 200 million text messages across the world each day. While such projects and programmes are ostensibly directed towards terrorists or other 'enemies of the state', the difficulty of knowing who is posing or might pose a threat has resulted in a sweep of suspicion so broad as to include almost everyone and to exclude, in principle, no one.

The prevalence and level at which such surveillance operates has provoked much controversy. In 2014 there was a high-profile scandal when it became public knowledge that America was eavesdropping on the telephone conversations of its allies as well as its enemies. More generally, civil rights campaigners have objected to the loss of privacy associated with these levels of surveillance. Irrespective of their geographical location or national identity, each time someone logs on to the Internet, makes a bank transaction, uses a credit card, travels across a border, makes a telephone call, or files a tax return, they are potentially adding to an electronic trail of activities that provides a digital record of their bodily existence and activities.

The routine monitoring of data may never become more than that for the vast majority of people—with growing numbers becoming accustomed to their bodies being scrutinized through such methods as retinal scanning in airports, ID card entry to schools, and even hand geometry scanning in day-care centres. Those identified through surveillance activities as threats or potential threats to security, though, can find themselves the recipients of 'heavy', coercive forms of control. The rendition practices made infamous by the 'War on Terror', for example, involved suspects being taken forcibly to nations that permitted forms of interrogation legally forbidden within their country of domicile. This situation assumed an even more disturbing aspect with the publication of the 2014 Senate report on CIA interrogation/torture techniques—techniques that inscribed themselves directly onto the bodies of suspects, inflicting extreme physical stress and mental torment. While surveillance increasingly encompasses us all, then, different categories of people are affected by it in very different ways.

The contrasting ways in which surveillance affects people classified differently is also evident in terms of those social groups who find themselves targeted for monitoring. The sheer volume of data collected by national security agencies is too vast to be sorted manually, and surveillance operations frequently depend on software and recognition devices to highlight risks. Yet the human decisions informing these programmes are not neutral, concentrating their gaze on particular groups (as opposed to actions). In the case of closed circuit television (CCTV), for example, research in the UK and USA suggests that the criteria informing selection targets focuses disproportionately on black males.

The surveillance of bodies is not in itself new: as Foucault noted, it has long constituted a method of 'normalization' in prisons, schools, and factories. It is also worth noting that the most ambitious and extensive system of surveillance to exist in the second half of the 20th century was not to be found in capitalist America but in communist East Germany. Here, the state apparatus aimed not only to stop subversive acts, but also to exert control over the micro-level of people's activities, thoughts, and desires in order to prevent dissent or disorder from ever occurring. Yet while surveillance operations within the capitalist West may not have been as extensive at that time, they have undergone unprecedented growth since the collapse of communism.

In Britain, there was a large increase in CCTV cameras during the 1990s as a response to the mainland bombing campaigns of the Irish Republican Army. The focus here was on observing the movement of bodies through urban spaces in an attempt to identify suspicious individuals and packages. This was complemented by the use of informants and undercover agents, a tactic deployed increasingly in relation to *any* group considered to be an 'enemy of the state'. One recent case in the UK, for example, involved an undercover police officer who lived for years as partner with a woman he was investigating. While police representatives portrayed him as a 'rogue officer', it became clear that this was a textbook surveillance method resulting previously in officers fathering children with those they were investigating. Contributing to the reproduction of the population while seeking to regulate the actions of particular groups may be rare, but it illustrates the unintended consequences that can follow from certain attempts to solve what Turner identifies as the core body problems associated with governance.

While undercover infiltration constitutes a means of surveillance dependent upon agents crossing over the borders that separate 'legitimate' from 'illegitimate' members of a nation, most surveillance operations seek to strengthen the boundaries that exist between these bodies. This is evident in those passports, identity cards, iris, fingerprint, and voice recognition devices, and the gathering of other bodily data to scrutinize and monitor physical identity, all of which are now staple features of border control. One of the first uses of biometric data to maintain borders occurred during the early 1980s across the Mexican-American border in an attempt to reduce the flow of drugs. The European Union also developed biometrics systems in the late 1990s in order to allow asylum seekers to be identified and authenticated. As the cultural studies scholar Ann Davis suggests, the body is here treated as a password.

If surveillance operates to highlight and help enforce the boundaries between legitimate and illegitimate bodies, it also does the same for 'acceptable' and 'unacceptable' places and spaces embodied subjects might visit, including those involving the borders of the Internet. In China, for example, the Golden Shield Project censors the web content that can be accessed by individuals, while North Korea has adopted an even more restrictive policy towards Internet access. Most liberal democracies proscribe as illegal certain violent and/or pornographic content on the Internet, while abusive or threatening messages can also fall foul of the law. All these means of scrutinizing and proscribing mediated bodily movements and activities—whether we evaluate them as oppressive or not—require high levels of deliberation, reflexivity, and planning on the part of state securities. They involve embodied subjects being conceptualized and apprehended in particular ways, and assuming a specific value for those who have attained the power to govern them.

It would be wrong, however, to think that surveillance is the exclusive prerogative of central governance. Covert filming is used by parents as a means of collecting evidence about the quality of care afforded to their children by child minders, for example, and as a way for adults to check on the treatment of elderly relatives in care homes. Helmet cameras are becoming increasingly popular among cyclists as a way of recording evidence of dangerous driving. Elsewhere, smart phones, iPads, and other computer tablets are routinely used by protesters to record and prevent police intimidation and violence.

Away from such instances of 'counter-surveillance' is a growing trend towards *self*-surveillance. Dieting involves counting calories and monitoring the self through the ritual activity of recording weight (sometimes in dedicated 'weight watchers' groups). Exercise has become a carefully monitored activity, far removed from the uncertainties and vagaries of play: it is assisted by apps that measure distance travelled and calories burned, workout schedules that count progress in terms of sets and repetitions, and pedometers used by participants in 'older adult fitness walking'. These are not imposed by others, but are embraced by individuals keen to place their bodies under scrutiny, and to measure their own progress in relation to the norms of their peer group or other population. Such self-surveillance has reached its most developed expression in the 'quantified-self' movement. This consists of a growing numbers of individuals dedicated to recording their lives through technologies that track everything from their mood to their food consumption and their performance across a range of daily activities.

If individuals have employed self-scrutiny as a means of disciplining their own bodily desires and actions, surveillance has also increasingly become a form of entertainment. We may no longer flock to see public executions—although the videos posted by ISIS and other extreme Islamic groups still make it possible to witness such spectacles—but reality television programmes such as *Big Brother*, *Survivor*, *America's Next Top Model*, and *Nanny 911* have become staple ingredients in television network schedules over the past few decades. These provide viewers with access to the actions, conversations, and humiliations of participants. They add to the sense that surveillance is not the prerogative of central governance but appears to have been adopted by society as a whole: the experiences, expressions, and actions of embodied subjects have never before been so extensively and intensively recorded and observed.

Governing life's processes

Focusing upon the body in terms of the data or information that can be extracted from it in the case of security surveillance can be associated with governmental techniques that facilitate 'heavy' interventions into suspect individuals, as well as 'softer' and more 'productive' techniques as evident in the case of the quantified-self movement. For the sociologist Nikolas Rose, however, the latter tendency to scrutinize the smallest details of bodily presence and activity also signifies a broader shift in the level that *medical* governance has come to operate upon the body.

Rose acknowledges that the body 'as a whole'—the focus of clinical medicine in the 19th and early 20th century—may remain the basis on which most people tend to experience, imagine, and act upon their bodies. Nevertheless, complementing the scrutiny of data trails left by the smallest technologically mediated actions of individuals, there has also been a related micro-level concern with the capacity of science to manage, engineer, and regulate 'life itself'. Thus, biological advances in research, following the mapping of the human genome, have increasingly promoted a view of embodiment as something that can be controlled through the monitoring and manipulation of its most basic processes. Being able to identify what are known as single base variations in the genetic code of individuals (the small differences that exist between people's DNA), for example, holds out the promise of assessing an individual's risk in relation to cancers and cardiovascular disease and of developing medicines that are thoroughly personalized.

This active approach to managing the micro-processes of our biology is reflected in contemporary research within neuroscience that suggests the 'plasticity' of the brain makes it possible for individuals to increase their creativity and concentration and to avoid negative thinking that might result in depression or a lack of productivity in their social and work roles. Genetic or neurological risks are no longer matters to be responded to through 'heavy' eugenic policies, but matters for constructive and 'positive' intervention through gene therapy, cognitive training, and general lifestyle changes.

Such opportunities may promise historically unprecedented degrees of self-control, but they also place on embodied subjects a considerable burden of responsibility and self-governance. The recent growth of notions of biological and neurological citizenship, for example, contain the implication that people need to monitor, evaluate, and work on themselves using as their guide expert knowledge from the 'received facts' of science and medicine. Knowing one's chances of inheriting particular diseases, as well as being aware of the risks associated with lifestyle choices related to diet, exercise, and alcohol, can prompt individuals to become what the anthropologist Kaushik Sunder Rajan refers to as 'patients in waiting'. The patient in waiting governs him/herself on the basis of a medical paradigm that highlights the economic value of physical and mental health to the individual, to the health service, and to national productivity.

If these notions of medical self-governance become more prevalent, they pose particular challenges for those growing old at a time when ageing too far from the ideal of youth is seen as being akin to a disease. They also raise concerns about the treatment of individuals who struggle to cope with significant physical and mental disabilities, and who simply cannot manage their condition without regular help from others. Elsewhere, particular pressure is already being placed on pregnant women who are increasingly seen as the agents of 'responsible choices' in the realm of reproductive genetics. Making a conscious decision not to use prenatal testing for abnormalities is very different from not having the technology that would allow for such testing in the first place, and issues of liability loom large if such a decision results in a child being born with serious disabilities. The disability studies scholar Tom Shakespeare suggests that this situation has been associated with the notion of 'irresponsible parental choice' and argues that its consequences can be viewed as a form of 'weak eugenics'. More widely, conceptions of 'biological citizenship' and the 'patient in waiting' place pressure ultimately on each and every embodied individual. Irrespective of how strong, fit, healthy, or autonomous individuals may feel they are, our very existence as interdependent physical beings confronts everyone who lives long enough with the limitations and frailties associated with being human.

The self-governance that has been associated with this focus on the micro-processes of life itself may be a 'positive' form of control in Foucault's terms, but it also raises the possibility of states retreating further from principles of universal health provision and of people becoming preoccupied increasingly with 'bare life'. The capacity for individuals and organizations to glean knowledge at the micro-level of embodiment also has implications for health insurance and for social policy. If insurers are allowed access to the biological and neurological profiles of individuals, there exists the possibility that individuals will be refused cover or allowed it only on the most unequal of terms. This is a contentious issue in health systems across continents. In a related move, the genetic testing of employees and potential employees gathered pace rapidly in the USA during the 1980s. While in 1992 Wisconsin became the first state to outlaw genetic discrimination in employment and insurance, the future acceptance of medical self-governance could well see these issues opened up to renewed scrutiny.

Finally, while scholars such as Rose focus on how increasing knowledge of the micro-processes of life may contribute to self-governance, the sociologist and bioethicist Troy Duster has demonstrated how genomic research has been accompanied by a tendency for scientists to engage in the logically dubious practice of associating molecular inheritance with racial ancestry. Current technology is unable to trace significant parts of our genetic make-up. Despite this, however, scientists have frequently reintroduced crude racial categories to assessments of the molecular composition of human beings. The problem with this is that it threatens to create new forms of racism that may resurrect previous justifications for social inequalities based upon bodily differences.

The state, life, and security

The surveillance of whole bodies and data trails for the purpose of security, and growing attempts to control the processes of life at the molecular level, may appear to represent radically different levels of governance. While the former has been associated with the protection of post-colonial privilege in the case of the USA and Europe, through the establishment of 'fortress' continents, the latter has (at least from the most optimistic of perspectives) been associated with a politics of hope for humankind.

During the last two decades, however, the spectre of terrorist attacks that utilize biological agents has brought these forms of biopolitics closer together. This is exemplified by US defence policy. Shortly after America's launch of the 'War on Terror', and the anthrax attacks that followed 9/11, the Bush administration initiated a national defence strategy against biological threats. Concerned with protecting people from threats at the molecular level, the US Congress approved in the same year unprecedented levels of funding for storing vaccines. US Centres for Disease Control are concerned not only with general issues regarding public health, but with identifying lists of bioterrorism agents, while the Pentagon's Defense Advanced Research Projects Agency (DARPA) views infectious disease in terms of its potential to be a security threat.

As the science and technology studies scholar Melinda Cooper argues, the 'War on Terror' has taken a biological turn, and the molecular building blocks previously identified as facilitating the health and revitalization of embodied life are now also viewed as possessing the potential to threaten and destroy individuals and populations. Ironically, we see here the turning of micro-level biological processes against the embodied individual as a whole. The fear that dangerous biological agents could be introduced invisibly across otherwise apparently secure borders, to mutate across species and infect humans, has led to a near dissolving of the boundaries between public health and bioterrorism. For the Italian political philosopher Giorgio Agamben, this development can be seen as another step in the reduction of biopolitics to issues of bare life—the basic processes of life itself.

Biopolitics is alive and well in the current age of governance, but the conception of the body with which it is associated has undergone significant change. The medieval focus on the dead body of the criminal was replaced by an early modern concern with the management of differentiated populations. However, the significance of tracing and tracking data trails, and seeking to manipulate the body at its molecular level, has resulted in a further shift to what is involved in managing the 'problem' of the body. The contrasting forms of control that have been exerted over embodied subjects bring into view very different ways of understanding and valuing the body, and serve to complicate further any straightforward answer to the question 'What is the body?'

Chapter 5
Bodies as commodities

Bodies have been conceptualized and valued in a wide variety of ways across contrasting cultures. Historically, for example, they have been viewed as resources to be owned and traded since the development of slavery in ancient civilizations across the world. Over the last few decades, however, there has been a multiplication of methods through which the physical appearance, organs, and flesh of the living and the dead have become implicated in market transactions.

In seeking to understand why these processes have become increasingly pervasive, the writings of the radical philosopher and economist Karl Marx are useful. Marx insisted that capitalism required a class of formally *free* labourers, able to sell their labour to whichever employer they wished. He also argued that in societies dominated by the profit motive, people's appreciation of their senses, of other people, and of the world around them revolved around issues of who *owned* what. These tendencies were exacerbated by economic crises that exerted downward pressures on wages, and reduced ever more human life and nature to mere means for the production of value.

Such conditions help to explain why the distinction between embodied subjects and commodities has become increasingly blurred. By the early 21st century, indeed, the consequences of bodies being subjected to market processes were experienced globally with increasing intensity across all sections of society. Among the more privileged, this commodification was evident in the various ways people sought voluntarily to cultivate their appearance to enhance their standing in the marketplace of work or personal relationships. The flesh here becomes a form of physical capital that can be utilized for economic benefit. Among those dispossessed from the benefits of global capitalism, however, the commodification of bodies has also involved the brutal and coercive selling of women and children into the sex industry. In these cases, embodied subjects are reduced to enslaved resources from which can be generated future profits.

If embodied individuals as a whole are sometimes reduced to the status of commodities, developments in science have also facilitated a massive expansion in the international trade in body *parts* and *processes*. Organ trafficking has become an international problem, while the biotechnological exploitation of DNA has been associated with potential medical advances that have attracted billions of dollars of capital investment. Undertaken voluntarily or involuntarily, involving the individual as a whole or in part, bodies have never been so multiply or thoroughly entangled in circuits of financial value.

The increasing commercialization of our bodily selves as a consequence of contemporary economic developments can be viewed as eroding or transforming what it means to be an embodied subject, highlighting the uncertainty over what constitutes a body. Nevertheless, the various ways in which bodies have been seen as objects and resources to be transacted for profit have not occurred without opposition or counter-developments associated with very different ways of valuing the body.

Marketing appearance

By the start of the 21st century it had become routine for multinational corporations and local businesses alike to require their employees to embody a particular look expressive of a company brand or image. Minimally, this involves being required to dress in a particular way and to project particular emotional impressions—associated variously with images of authority, competence, cheerfulness, and caring—to colleagues, clients, and customers. Such requirements are hardly new (having, since the second half of the 20th century, been standard for a range of occupations in the service sector), though they have extended in recent years to minimum-wage jobs in coffee outlets and other retail centres on the high street.

The quantity and quality of this bodywork and emotion work required of employees in recent decades has become a prominent issue in academic debates. This is reflected in the sociologist and columnist Barbara Ehrenreich's (1989) book, *Fear of Falling*, on professional life in America. Ehrenreich detailed the increasing emphasis placed on having a marketable body in the labour market. She also suggested that this development had prompted the professional middle classes to reflect on and scrutinize their bodies in relation to the ascetic demands of the workplace; developing (through exercise, diet, and an avoidance of heavy drinking and smoking) a form of physical discipline associated with maintaining a youthful appearance that gives them an advantage over their peers.

Such fears of appearing insufficiently youthful to maintain one's working status appear if anything to have become more acute since Ehrenreich's observations. In Silicon Valley, for example, business is thriving for plastic surgeons who report a growing use of Botox as well as other 'age denying' procedures among male tech workers in their thirties. Women have long been subject to such pressures, and have for decades harnessed cosmetic techniques and operations in attempting to re-enter or stay in the workforce, yet these demands have intensified along with an extension in the technological means available to satisfy them. In California, complaints against age discrimination are reported to be higher than those associated with discrimination on the grounds of race or sexual orientation. Being caught in possession of a body that has a declining market value can indeed result in the loss of employment.

Undertaking bodywork in order to maintain or gain a marketable appearance is not confined to America or Europe but has become increasingly global. Research on aesthetic surgery in South Korea, for example, has documented particularly high rates of invasive treatment among men and women. In 2008, at least 20 per cent of Koreans underwent some form of cosmetic treatment. Far from simply being an attempt to look more Western, this bodywork is directed towards cultivating what is considered to be an idealized 'natural' Korean look reflective of youth and high social standing. In a country that has high rates of further and higher education, such a look can be an important variable in securing a good job.

The marketable value of appearance is not only confined to the workforce in Korea, it also extends to Koreans' success or failure in the field of intimate relationships and marriage. Here, the face and body become markers of value that can help individuals increase their chances of gaining a successful and attractive partner. Once secured, moreover, such a partner may also prove to be a status symbol and an asset in navigating the social elements associated with gaining popularity and promotion within an organization.

Commenting upon the growing general significance of appearance as a bearer and vehicle of value within capitalist societies, the French sociologist Pierre Bourdieu suggests that the cultivation of bearing, taste, manner, and speech have become widespread markers of social class and status. Bourdieu focuses predominantly on how the body enters into what he refers to as the 'pursuit for distinction' in France, but his account firmly establishes its significance for people's chances of success within *any* market based society. For Bourdieu, indeed, visual appearance and other impressions 'given off' by the body are critical for people's capacity to accumulate value *across* the various dimensions of social and economic life. In this context, it seems to be the case that if people are increasingly treating their bodies as projects, as suggested in previous chapters, they are encouraged to do so in relation to the growing marketization of life. While appearances have become subject increasingly to these market based pressures, however, so too have other aspects of embodiment.

Medicalizing bodies for profit

The preoccupation with appearance as a means of enhancing an individual's value in work, marriage, and other areas of life has grown in recent decades, especially among those possessed of the means to treat their bodies as projects. However, this period has also given rise to scientific advances into the building blocks of life itself. Such developments have stimulated far deeper commercially driven interventions into the molecular level of people's bodily being.

In the late 1970s and early 1980s, advances in recombinant DNA technology—which facilitated the isolation and combining of DNA molecules in labs, enabling scientists to edit, recombine, transplant, and produce genetic and other living material—stimulated the development of the biotechnology industry. Symbolized by the public launch of the company Genotech on Wall Street in 1980, the sociologist Catherine Waldby has suggested that such events reflect the growing significance of *bio value*. Interpreted broadly, bio value refers to processes that enable bodily material to be exploited for the development of medical and other products.

It is in medicine that this particular commercialization of bodies has most potential for profitability. The minute variations that exist between people's genetic make-up means that some individuals are more vulnerable than others to particular diseases such as lung cancer or cardiovascular disorders, and the promise of what is termed 'pharmacogenomics' research into these variables has been associated with a potential revolution in health care. The promise of such developments was such that capital investment in these new technologies grew rapidly in the 1990s.

Biotechnological research directed towards future innovations is not, however, the only way in which our bodies are being 'mined' for value in the area of medicine. In his 2012 book, *Drugs for Life*, the medical anthropologist Joseph Dumit notes that the average American is prescribed between nine and thirteen prescription drugs per year, and that this amounted to over four billion prescriptions in 2011 alone. For the pharmaceutical companies, patients who are not taking medicines constitute 'prescription loss'.

In this context, the pharmaceutical industry's attempts to persuade individuals to spend money on what Dumit calls 'surplus health' becomes an important means of maintaining and enhancing profitability. In areas as diverse as sexual performance, emotional mood, and weight loss, multinational companies bring new drugs to market and invest large quantities of money in advertising them to doctors and selling them to the 'worried well' and those seeking to 'live life to the full'. Everyday and unremarkable levels of tiredness and unhappiness have been turned into sexual dysfunctions and depressive disorders amenable to medical treatment, while excess flab and even hunger itself become indicators of actual or potential obesity to be treated by appetite reduction drugs.

Research into the development of new medicines, and the continued exploitation of existing medicines to a growing global market, raises the question of how the information provided by our DNA and our genetic make-up can be 'owned' in order for profits to be made. The bringing to market of new medicines requires significant investment if experiments are to be turned into trials and then into products, and the existence of patents to protect inventions and developments is crucial to this process. Thomas Jefferson introduced America's first patent act in 1793, enabling profits to be placed on a secure legal footing for the invention and development of machinery. More generally, and far more recently, the agreement of Trade Related Aspects of Intellectual Property Rights (TRIPS) was the most comprehensive international intellectual property treaty of the 20th century. Initiated in 1994 and administered by the World Trade Organization, TRIPS provides a global basis on which companies can pursue the profitability of new products, including medicines.

Such international agreements have encouraged companies and individuals to seek to patent the products of biotechnological research involving DNA and, indeed, the human genome. As bodies become 'knowable' and can be understood on the basis of information and formulae relevant to DNA, the ownership of patents covering intellectual work raises questions about the proprietorship and commercialization of bodily life processes. In 1991 and even before the instigation of the TRIPS agreement, for example, the scientist and entrepreneur Craig Venter sought patent protection for genes to be found in the human brain. Since then, various legal systems and states have been involved in ongoing debates and decisions regarding how far it is possible to view the building blocks of human life itself as commodities that can be owned and exploited for profit.

These attempts to exploit commercially basic life processes raise issues not only about the commodification of bodies in general, but also about the global winners and losers in such developments. Certain nation-states have insisted on commercial agreements before allowing the exploitation of DNA samples and genetic research, or have prevented this from happening at all in an assertion of what has been referred to as 'biosovereignty'.

Despite such assertions of national control, however, multinational companies dominate the production of exploitable knowledge at the molecular level of the human being. This often involves biotech researchers collecting samples and data from local populations across the world even though many of these individuals will never be able to afford the medicines produced from such research. Similarly, the risks faced by those participating in clinical trials are not spread proportionately among rich and poor. Clinical trials have been outsourced from the USA, especially from the mid-1990s, to India and other countries disadvantaged by the inequities embedded in patterns of global economic development.

Trafficking body parts

Concerns about the relationship between global inequalities and the medical commercialization of bodies have not been confined to clinical trials and testing.

In 2011 the academic journal *Body & Society* published a double issue that focused on what Nancy Scheper-Hughes referred to as 'transplant tourism' or 'transplant trafficking'. Illegal though many of these practices are, they constitute 'a global billion-dollar criminal industry involved in the transfer of fresh kidneys (and half-livers) from living and dead providers to the seriously... ill and affluent or medically insured mobile transplant patients.'

The historical antecedents of transplant trafficking include the 19th-century grave robbers in the USA and UK who sold corpses illegally to medical schools for the purpose of dissection. More recently, there have been cases ranging from the illegal harvesting of skin and organs from corpses in New York funeral homes, to the kidnapping of Mexican children for the purpose of using their organs for the transplant needs of American children. This latter case, one of the most notorious of its kind, involved clinics on the USA/Mexico border as well as dozens of Mexican medics and other professional accomplices.

The trade in body parts is not confined to a few countries. Organized 'transplant tourism' developed in the Middle East during the 1970s when patients travelled to India and then (after problems arose with infected and diseased organs) to private hospitals in the Philippines to buy kidneys. In Europe, Moldova and the Ukraine are currently home to large numbers of donors willing to sell their organs, while Turkey is the country where most clandestine operations are conducted. Elsewhere, Brazilian kidney sellers travel to South Africa to do business with Israelis, while the current levels of kidney selling among certain areas of India is such that one survey reported up to 10 per cent of households in particular localities contained one member who had sold a kidney.

Transplant trafficking depends upon and is made possible by abject poverty. The vast majority of people who sell one of their kidneys may formally do so through choice, but their decisions are made in the context of seeking to provide a tolerable standard of living for themselves and their families. In Moldova, for instance, approximately one-quarter of its four million population have left the country in an attempt to find work. Circumstances such as these provide intermediaries or 'organ brokers' with a fertile context in which to match ready sellers with wealthy foreign patients. Decisions to sell organs are not only made in conditions of great poverty, however, but also often accompanied by ignorance about the post-operative problems and long-term effects on health that can follow for donors unable to access quality medical care.

The organ trade also sometimes has a more directly coercive element to it. As the ethnologist Susanne Lundin reports, there have been cases of Moldovan men travelling to Istanbul with the assistance of agents who promised them a job. On arrival, they discover that the prospect of work has vanished, and they are kept forcibly locked up in accommodation until they repay the agent for expenses incurred. The only alternative to money, which they do not have, is payment in the form of a kidney.

The illicit business of organ trafficking is part of a much larger commercially driven set of medical migrations. These are often considered perfectly legitimate ways of securing medicines and medical treatment that would be unavailable or unaffordable to individuals in their own country. Certain states, moreover, have actively promoted medical services to foreign patients as part of a broader economic strategy. In Thailand, for example, medical tourism provides a significant source of revenue for hospitals and the tourist industry, as well as a taxation stream for the state, with over a million foreigners arriving for treatment each year. Nevertheless, such strategies also have implications for the health care made available to indigenous populations. In South Africa, for example, expensive medical operations such as organ transplantation are now confined to the private sector—a sector that also employs significant numbers of highly trained surgeons no longer working for the majority of the population.

The unequal availability of health care entailed by these developments has been referred to as a body tax on the world's poor, but the appropriation of the organs of the poor in order to facilitate the health of the relatively wealthy takes this 'tax' to a whole new level. As if this is not disturbing enough, there are also in existence international circuits of trade in which embodied individuals as a whole are sold to others in forms of modern-day slavery.

Enslaving bodies

The geographer David Harvey has suggested that we are living through a period of 'primitive accumulation' reminiscent of the conditions that first gave rise to capitalism. This accumulation involves obtaining wealth and restoring profitability through force and domination: contemporary examples include organized crime, financial fraud, and market manipulation that bankrupt small businesses and damage savers, and international financing arrangements that leave nations struggling with unmanageable debts.

When these circumstances exist for any prolonged period, they can prompt the breakdown of normal capitalist wage relations. One consequence of this is the growth of informal and undocumented migrant labour: individuals working abroad in dirty and dangerous jobs without rights and receiving remuneration below minimum-wage levels. Segments of the catering and agricultural industries in Europe, and significant sections of the agricultural, cleaning, and construction workforce in the USA are just two sectors dependent on this exploited workforce.

The breakdown in formalized wage-relations associated with the return of conditions approximating to primitive accumulation has been accompanied by a rise in human trafficking and the systematic use of forced labour. Figures cited by the legal consultant Penelope McRedmond suggest that 2.5 million people are trafficked each year into various forms of modern-day slavery in which bodies are sold and bought for the purposes of involuntary sexual or labour exploitation. Much trafficking occurs through clandestine methods, but it can also take place in heavily policed public areas. In 2006, for example, auctions involving young women destined for sexual slavery took place in front of a café at Gatwick airport, London.

Trafficking involves coercion through violence and the threat of violence, blackmail, and psychological domination, and has contributed to a situation in which over twenty million people are reportedly trapped within various forms of slavery. While the International Labour Organization (ILO) puts the figures at twenty-one million, the 2014 Global Slavery Index (the main report of the Walk Free Foundation) puts the number as high as 35.8 million. The illegal and 'underground' character of much slavery makes it difficult to get precise figures. What we do know is that India, China, Pakistan, and Nigeria contain the largest numbers of slaves, while slavery remains a structurally intransigent part of other societies. In Mauritania, for example, ethnic differences have served as a basis for the longstanding enslavement of 'black moors' by Berber Arabs.

The rewards are high for those prepared to trade people in these ways: the ILO has estimated that the profits from this (in)human trade total $150 billion per annum. The problem has become so serious and widespread, indeed, that each year since 2001 the US Department of State has released a global 'Trafficking in Persons Report' designed as a comprehensive resource to track and inform policies and practices to help combat the problem.

Modern-day trafficking is a truly global phenomenon. In America, for example, instances of trafficking and coercion have ranged from the purchasing of Cambodian babies for adoption (a practice that frequently involved intermediaries buying and even stealing babies from their families) to the control and tattooing of prostitutes by pimps. Elsewhere, its various forms involve the selling of young women and children from Burma, Mongolia, Thailand, and elsewhere, as brides in China. It also includes the hundreds of thousands of women trafficked each year into the European Union, from Eastern Europe, to be sold in the sex industry; and the North Koreans working as 'state-sponsored' slaves in the construction industry within Qatar.

In the Middle East more generally, forced labour extends to the *kafala* system of migrant work. This plays an important role in the construction industry and in domestic labour by tying migrant workers to sponsoring employers, often removing from them the papers they need to return home, and frequently leaving these individuals in debt by charging them more for shelter and subsistence than they receive in wages. Elsewhere, children and adults are forced to harvest cotton for certain periods by the government in Uzbekistan, while children are subjected to slavery in diamond mines in the Democratic Republic of the Congo, and are forced to fight in wars in Sierra Leone, Sudan, Afghanistan, and elsewhere. No country seems immune from the problem.

Global in scope, the enslavement of embodied subjects occurs against a background of longstanding patterns of post-colonial exploitation and other international patterns of dependency, inequality, and political instability. The continued existence of US military bases in such countries as South Korea has been identified as increasing the trafficking of women into the sex industry, for example, while the spread of market forces in the former communist bloc has prompted a mass growth in forced labour and prostitution. Newspaper adverts in Russia, Ukraine, and elsewhere promise work abroad in childcare and waitressing, only to condemn those hoping to escape poverty to enforced servitude.

Those involved directly in trafficking and slavery may receive one-off payments for selling individuals to others: the enslaved individual here is reduced to the status of a profitable commodity. In other cases, however, those who benefit repeatedly from the value created by those forced to work in the sex industry, construction industry, or in any other type of enslavement are effectively treating those coerced as a form of fixed corporeal capital that can be used to generate repeated profits in the future. As Marx argued, if a slave owner 'loses his slave, he loses his capital'.

In terms of the deep backdrop for these developments, the widespread commercialization of sex and what has been referred to as the 'pornification' of culture has stimulated an attitude among many that it is acceptable to buy sexual services. Trafficking and slavery only exist because of demand for the goods and services they satisfy, and there is no sign of a declining appetite for cheap products or the willingness to satisfy all manner of desires through legal or illegal transactions (see Figure 9).

It is also important to note that there is a significant cross-over between the illegal profits earned from slavery and apparently legitimate businesses. As Louise Shelley (Director of the Terrorism, Transnational Crime and Corruption Center (TraCCC)

9. The commodification of sex has intensified in recent decades.

at George Mason University) argues, examples of this include the Japanese Yakuza who use profits from such activities to invest in golf resorts in Thailand, and then make further profit from marketing vacations that include sex tourism.

Resisting commodification

Marketing of appearance, medicalizing bodies for profit, trafficking body parts, and selling embodied subjects into forced labour and slavery each highlight a breakdown of the distinction between physical subjects on the one hand and commodities on the other. It would be a mistake, however, to think that the commercialization of bodies and their transformation into distinctive forms of physical capital has been unopposed.

There are a large number of international bodies and anti-slavery organizations—international and national, secular and religious—working to end forced and indentured labour, and an increasing number of countries are legislating against modern forms of slavery. The commodification of body parts, fluids, and processes has also been challenged by alternative practices. The World Health Organization (WHO) reported an extra 8.6 million blood donations from unpaid volunteers from 2004 to 2012, for example, with the highest increases coming in Africa and South East Asia. The WHO also report that a majority of countries they have data on collect over 90 per cent of their blood supply from unpaid volunteers.

If blood remains implicated significantly in a 'gift network' rather than a cash nexus, there also exist related practices in the field of organ donation. In the USA, for example, the 'kidney daisy chain' has been viewed as an instance of 'new donor ethics' in which donors who proved incompatible with their intended recipient donate to someone else in the hope that this practice will become generalized.

The public health expert Klaus Hoeyer and his colleagues account for these developments by arguing that the extension of markets has developed *alongside* an extension of notions of embodied personhood. This can be seen across a number of different countries in relation to a variety of issues. In Denmark, for example, embryonic cells were not regulated directly before the mid-1980s but have featured increasingly in debates about what constitutes a person. Elsewhere, in Alder Hey Children's Hospital in England, there was an inquiry in 1999 into the unethical retention of organs and tissues from dead children, a retention that did not involve the consent of parents. This prompted a public and media response informed by a conviction that the body and its organs retained links to the identity of the children of which they were a part and could not be reduced to mere objects post-mortem.

More generally still, contrasting definitions of life and death continue to limit organ donation in countries such as Japan. Religious issues regarding the importance of not violating the integrity of the body borrowed from God also remain influential in Muslim countries such as Turkey. Elsewhere, it is worth noting that the acceptance of brain cessation as the marker of death in America—a position that facilitated increased organ donations—took place only after significant debate and legislation.

In addition, the experiences of transplant recipients continue to suggest that benefiting from this exchange cannot be viewed in the same terms as buying other commodities. Various anthropologists describe a belief that the 'cell memory' of transplanted organs influences the personality of the recipient as a form of resistance to the objectification of body parts. As Aslihan Sanal's study discovered, transplant recipients often found themselves feeling as if they were inhabiting different worlds, possessing part of another person's spirit that affected their sense of self.

Finally, there are also strong philosophical arguments opposing the commodification of body parts. Immanuel Kant associated human dignity with the integrity of the embodied subject as a whole. Vital organs are bound up with the constitution and development of who we are as embodied human subjects, and the capacity of individuals to act morally cannot be separated from our existence as beings possessed of minds and bodies. More significantly still, Kant placed at the heart of his philosophy the injunction that it was wrong to use people as mere means to ends. From this perspective, treating individuals or body parts as commodities for the realization of profits attacks the very basis of what it is to be a human being.

Whether or not the developments explored in this chapter do erode the basis of what it is to be human, however, there is no doubt that the commodification of embodied persons is continuing apace in the current era. Having introduced these discussions with reference to Karl Marx's insights into why it is that the distinction between bodies and commodities can become blurred within capitalism, it is worth returning to his thoughts on the subject. For Marx, indeed, the primitive accumulation of which slavery forms a part 'is as much the pivot upon which our present-day industrialism turns as are machinery, credit, etc.' The enslavement of bodies enabled colonial powers to gain value from cotton and other raw materials that accelerated world trade and formed a basis for the development of modern industry. Contemporarily, forced labour exists alongside the commercialization of appearances, of body parts, and of medical research and products involving life's basic processes. Bodies are being exploited in multiple and often novel ways as sources of economic value in their own right and as vehicles for the creation of profit.

Chapter 6
Bodies matter: dilemmas and controversies

Previous chapters have explored how 'starting from the body' can provide us with the basis of a novel and productive approach towards the analysis of society, culture, history, and identity. In so doing, they focused on highlighting alternatives to those Western traditions of thought that marginalized the physical dimensions of social and personal existence, elevating the mind over the senses and ignoring how human thought as well as action occurs through our embodied being. Opposing this tendency, I argued in these discussions that it is vital to recognize that bodies matter—possessing their own properties that change over time while simultaneously being permeated by and situated within a wider social and material environment. People's capacities to make a difference to social life exist because they actively 'imprint' their bodies on to and mould the environment around them, while institutions and customs shape each new generation by sanctioning the cultivation of people's embodied capacities in certain directions rather than others.

This need to develop a dynamic approach towards embodiment matters has been reflected in the three themes that have permeated the various chapters. These have highlighted: (1) the capacity of social and technological forces to inform, and change, what has been understood conventionally as the biological constitution of our embodied being; (2) the uncertainties raised by this changeability about what our bodies are, and how they should be managed; and (3) the contrasting and contested ways in which bodies and embodied subjects have been valued.

In order to explore further the contemporary significance of these themes, this final chapter will explore briefly how each of them is associated with an issue involving the body that looks set to become more rather than less important. It will do this by posing three key questions about contemporary embodied identities and the quality of social relationships more widely. These questions address respectively whether recent changes in our bodily capacities have been associated with a decline in the moral standards of communication and interaction with others; how it is that we manage our potentially changeable bodily identities; and whether contrasting conceptions of what is valued, prized, and even considered sacred about the body raise challenges for the future of humankind.

Are mediated bodies immoral?

Social and technological advances have long altered the capacities as well as the characteristics of embodiment, prompting changing answers to the question, 'What can bodies do?' The growing abilities of our prehistoric ancestors to control fire, for example, was one of the most significant technological innovations ever, allowing people to inhabit previously inhospitable environments. Fast-forward to the present day, however, and it is possible to argue that the development of digitally mediated communication (facilitated by networked computers and the inception of the Internet) has brought about an unprecedented change in how the capacities of embodied subjects have been supplemented and enhanced.

Social networking sites, video messaging, electronic mail, blogs, online gaming cultures, and digital distribution platforms such as YouTube are just some of the means through which it has become possible to engage in mediated communication. In exploring the embodied implications of this, the Canadian

Cultural Studies scholar Vince Miller argues that these technologies operate by distributing the presence of bodies over time and space. Crucially, he also suggests that this change has been associated with a worrying decline in the moral quality of communication.

In order to understand the background to Miller's argument, it is first important to say something about how the relationship between embodied interaction and moral behaviour has been understood conventionally. Of particular importance here is the contribution made by Erving Goffman, the sociologist. Goffman argues that routine face-to-face interaction imposes certain limitations on how people relate to each other—such as the need to be physically proximate and to engage in turn-taking—if communication is to be possible and successful. He also suggests that these same conditions help promote a moral order of interaction. This is in part because participants not only have to be available and 'open' to others (exposing themselves to a degree of vulnerability), but must also extend a degree of trust to those with whom they are engaged. Without being prepared to accept that this can be done in relative safety and to the potential satisfaction of all parties, the interactions necessary to embarking on any shared activity would become impossible. Physically co-present interaction, in short, imposes conditions on bodies that generally make it in people's interests to extend respect and goodwill.

If face-to-face, corporeally co-present communication usually requires a degree of mutual goodwill, Goffman also argues that the moral qualities of these situations are reinforced by the fact that those who betray this trust (through insensitivity, rudeness, or being exposed as tricksters or fraudsters) stand to become morally tainted and disreputable. When enough people know about their behaviour within a locality, indeed, they are likely to be excluded from this moral order of interaction. Networked digital communication technologies, however, institute radical changes to the conditions in which much interaction occurs.

In particular, these remove the necessity of face-to-face meetings in order to interact with work colleagues, friends, and strangers, and also frequently reduce the disincentives against and consequences of behaving badly to others.

This is the context in which Miller suggests that mediated presence weakens the conditions that encourage individuals to develop a sense of moral responsibility. Evidence for this can be found, he suggests, in the frequency of 'flaming' (posting or sending offensive messages to others) or 'trolling' (deliberately provoking arguments) on social media sites. Instead of dealing with imminent face-to-face interactions, individuals are here being rude and offensive in the knowledge that they may never suffer the consequences of their actions. The effects of such actions can, however, be exceptionally serious.

The suicide of 42-year-old social worker Simone Black on Christmas day in 2010, for example, was a thoroughly mediated act for those who responded to her Facebook announcement: 'took all my pills, be dead soon, bye bye everyone'. Simone had 1,082 supposed 'friends' on Facebook, but her message provoked mockery and scepticism. None of her contacts alerted the authorities and, despite several being within walking distance of her flat, no one called round to check on Simone. Seventeen hours later, after Simone's mother was alerted to her post, the police found her dead.

This is not an isolated example. In 2008 a teenager from Florida announced on an Internet chat room that he was going to kill himself. After being encouraged to 'go ahead' by several online spectators, 1,500 people watched this suicide. Similarly, a British man hung himself a year earlier live on a webcam in front of online spectators after being goaded into the act in another Internet chat room.

In the context of developments such as these, Miller and other commentators have drawn on the writings of political economist

Adam Smith and the philosophy of Emmanuel Levinas in seeking to specify further why it is that face-to-face relationships stimulate a mutuality often missing from mediated communication. Distributed presence, it seems, is not just a technological matter but also raises critical moral considerations.

A very different area in which mediated presence can be seen as raising moral questions about interaction involves technologies of warfare. During the Gulf War, American pilots remarked that bomb target finders made the whole process of destroying targets (buildings, but also people) appear like a video game. Relatedly, the increased use since the 1990s of remotely piloted aerial systems, or 'drones', has prompted worries that battlefield deployments of automated weapons would result in more indiscriminate killing and would be insensitive to the interactional prompts that enable a soldier to distinguish a combatant from a civilian.

Mediated interaction is not, of course, the only or even the deciding factor that determines issues of morality within social interaction. As the sociologist Norbert Elias makes clear in his writings on the processes of civilization, medieval people at times took great delight in torturing and witnessing the torture of others in situations of immediate co-presence. Recently, the American abuse of prisoners at Abu Graib and Guantanamo was undertaken in situations of physical co-presence. Nevertheless, the concern highlighted by those who have written on the spread of mediated communication is that the incentives to act morally provided by *most* contemporary routine daily face-to-face interactions are weakening.

How do people manage bodily change?

Bodily change is a constant feature of social history and individual lives, and societies have long sanctioned surgical and other interventions into people's physical being. Nevertheless, the speed with which modern science, technology, and medicine continue to develop—and the enormous implications these advances have for people's capacities to alter and extend their embodied capacities—raise questions about how both individuals and institutions manage the opportunities and threats associated with these possibilities.

In the case of individuals, the extent and speed with which bodily related changes are now made available has subjected what was once regarded as the 'natural' and taken-for-granted conditions of physical existence to unprecedented levels of deliberation. These circumstances do not mean that we are divided, mind and body, against ourselves. Neither do they provide a warrant for those who believe that it is purely our capacity to think that defines us as human. As the pragmatist philosopher John Dewey makes clear, it would be impossible for us as intelligent humans to survive unless we bracketed out hundreds of potential contingencies and possibilities each day, took many of life's parameters for granted, and acted 'as usual' towards much of what we do and know. Nevertheless, the present pace of change does make it increasingly difficult to rely on many of our taken-for-granted actions as guides for maintaining and developing our bodily identities. Instead of continuing to depend on them, indeed, many of these embodied habits are likely to be subjected to reflexive scrutiny: an assessment of the self by the self, engaging with the possibilities opened up by these developments.

Growing opportunities for change, and the increasingly deliberative responses these are likely to stimulate, are also hugely consequential for institutions and those whose identities are associated with them. This is especially the case for institutions rooted in relatively fixed rules and practices that have evolved gradually through the weight of tradition, a type usefully illustrated by the case of religion.

Institutionalized forms of religious practice, faith, and personhood are associated conventionally with cultivating embodied habits in line with prescribed beliefs, fixed moral codes of righteousness and purity, and recurring rituals encompassing speech, diet, dress, and appearance. Such routinized patterns of religious socialization were suited to societies marked by limited patterns of social change in which moral dilemmas tended to remain stable over time. In the current era, however, religious authorities and their followers have few options but to confront the rapid developments occurring in the economy and society that alter the terrain on which they operate, even if their response is to develop particularly modern forms of religious fundamentalism. The advent of satellite television, the Internet, and other forms of global communication, for example, have made it virtually impossible for even the most 'orthodox' religious leaders and followers to be unaware of the strength and arguments of secularism and alternative beliefs in other parts of the world. Such circumstances do not entail that religion is unable to flourish, but they do cast doubt on whether traditional religious habits can endure.

In this context, it is interesting that religious authorities have made renewed efforts to consciously cultivate bodily practices and experiences that can help stimulate, stabilize, and reinforce belief. Pentecostalism and other forms of charismatic Christianity have become increasingly influential, and their focus on religious affiliation and conversion involves highly deliberative attempts to 'open up' the body to the power of the Holy Spirit. Central to this is a reflexive interrogation by the faithful of every aspect of life, alongside techniques of prayer and pure living designed to guide this scrutiny. As part of this process, churches make carefully planned use of physical images and objects—that have been referred to as 'sensational forms'—that can assist in nurturing and modulating religious emotions, feelings, and thoughts.

Pentecostal attempts to embody the presence of God have a long history in Christianity, but what marks out current practice is the extent to which this is carefully planned, regulated, and reflected upon by the individuals and authorities involved. The body pedagogics of the Alpha Course discussed in Chapter 3, for example, are explicitly designed and carefully scrutinized to provide the experiences and knowledge considered most suited for stimulating the presence of the Holy Spirit in the initiate.

This reflexive approach towards stimulating religious bodily experience is not confined to Christianity. It has been noted that many young European Muslims celebrate their religious identities as *choices*, marking them off as having made commitments that are distinctive from the predominantly Christian and secular societies in which they live. Elsewhere, research has suggested that many young Muslims in Iran engage with their religion reflexively, distancing themselves from the Islam of their parents and grandparents as a result of adopting a constructively critical approach to developing 'purer' religious practices; an approach that has heighted worries about the spread of fundamentalism.

Relatedly, the anthropologist Saba Mahmood has demonstrated how Muslim women in the Egyptian Piety movement seek through reflexive deliberations to scrutinize and deepen their personal religious orientations and dispositions. These women were very conscious of adopting highly conservative forms of dress, for example, as a way of stimulating within themselves a sense of propriety and shame; disciplining their bodies and experiences in line with Islamic teachings. Cultivating Islamic 'virtue' through their dress and demeanour, they sought to assess every aspect of their bodily selves and expressions as part of a conscious determination to ensure that both exterior actions and interior self conform to Islamic norms (see Figure 10). This involved seeking to cope with those who placed them in situations hostile to the achievement of piety in their daily life, and the internal struggle or *Jihad* with their own bodily actions and desires in a world that encouraged them to behave in impious ways.

Both these examples—from Pentecostal Christianity and Islam—highlight the significance of reflexive scrutiny when it comes to managing the body in the context of the options and

10. Dress is, for many, an integral aspect of religious identity and has provoked political controversy in secular societies.

choices that now face religiously inclined individuals whatever their affiliation. This does not suggest that habits have become irrelevant to social and religious life, but it does emphasize how individuals seek consciously to mould, form, and reform their bodies and habits in line with ideas and commitments. Habits are increasingly subject to reflexive judgements, and are criticized and where necessary changed by individuals consciously assessing how they wish to develop their self-identities.

The reflexive assessment of bodily habits and actions is undoubtedly more common among some people than others. These variations depend upon many factors including education and having available those resources that facilitate body management and that therefore make such deliberation of direct personal relevance. It is not just in the case of religion, however, that the extent and pace of change in the modern world is making an unquestioned reliance on past habits increasingly difficult. In the case of health, for example, people now have more information (and more conflicting information) about what they should and should not consume than ever before. Even those who continue to drink and smoke heavily, and refuse to participate in the exercise regimes lauded by local and national health services, find it difficult not to reflect on their behaviour in relation to pervasive messages of self-maintenance and 'biological citizenship'. Such examples can be found across social life and reinforce the sense that people's assessments, appraisals, and other reflections on their corporeal constitution have become increasingly important means by which they navigate their way through their world. Body matters encompass the embodied mind as well as the physical flesh.

Have our bodies become sacred?

A recurring theme throughout the preceding chapters has been the contrasting ways in which bodies have been valued and prized. The importance placed on particular views of what is 'natural', 'normal', or desirable about bodies at times, indeed, warrants the conclusion that there may be something sacred about human embodiment. This is a point that was made by the classical sociologist Emile Durkheim, who argued that although the body may appear to be mundane, it was frequently the location and even the source of sacred values 'set apart' from, prized, and rendered exceptional vis-à-vis daily life.

If we consider the relevance of Durkheim's comments in the contemporary era, it is reasonable to suggest that bodies are now prized and even rendered sacred by different groups of people on the basis of such varied factors as their youth, their ethnicity, their governability, their skills and capacities, their value as commodities, and also because they adhere to religious conceptions of divinely commanded living. Such variations, however, also raise the possibility that there may exist disagreement and conflict over what is sacred about the embodied subject.

In outlining the broad terrain covered by these contrasting and potentially conflicting approaches to what is prized about bodies, it is important to recognize the co-existence of both religious and secular conceptions of the body as sacred. Beginning with the former, there is no singular religious conception of sacred bodies, as evidenced by the distinctive prescriptions and prohibitions placed on prayer, fasting, diet, and lifestyle by common Jewish, Christian, and Islamic practices. The social consequences of these sacred conceptions can also be quite different: the traditional Christian allowance of a secular sphere on Earth (as expressed in the biblical phrase 'Render therefore unto Caesar the things that are Caesar's'), for example, is very different from the Islamic imperative that all of culture and society be subjected to the will of Allah.

The secular validation of the body as an object of political governance, in contrast, has placed heightened value on managing the basic process of life itself in terms of such issues as health and reproduction. Different again is that secular commodification of bodies and bodily parts that prizes the utilization of embodied subjects as resources for the accumulation of profit. This is not only in terms of bodies as producers, moreover, but also in terms of the capacities of consuming bodies to facilitate a company's profitability. Multinational companies have devoted significant resources, for example, to research into how brands attract allegiance from consumers who experience their own physical identities as bound up with the purchase and use of particular objects. From this perspective, we can suggest that 'Nike towns', the ESPN Zone in Chicago, Apple stores, and even coffee outlets such as Starbucks promote ways of using their products that treat consumers' bodily identities as sacred vehicles through which profits can be maintained.

These contrasting conceptions of how the body is perceived and treated as sacred have very different social, economic, and religious consequences—illustrating how our embodied properties and capacities can be steered towards and harnessed to distinctive ends. Different consequences also tend to follow the actions of those who transgress these secular and religious orientations towards our embodied being. Stigma and shame are frequently attached to individuals who fail to discipline their bodies in line with the secular demands of being efficient producers and consumers, for example, as evidenced by recent UK welfare policy which has tended to view as 'scroungers' people unable to work full-time because of disabilities. More visible is that 'holy rage' visited by fundamentalist groups upon people who treat their own and other people's bodies in ways that are seen as violating sacred principles. Examples of this include the attacks on abortion clinics in the United States during the 1980s by the Christian fundamentalist group, the Army of God. More recently and more widespread is the violence engaged in by Islamic groups determined to defend their visions of sacredly ordained forms of education and dress, and to punish those who dare to engage in profanating representations. Ranging from the shooting of Malala Yousafzai by the Taliban because she attended school and spoke up for the rights of girls to be educated, to the terrorist attacks on the offices in Paris of *Charlie Hebdo* for pictorially depicting and ridiculing Mohammed, the defence of sacred principles can be deadly.

Expansive bodies

Each of these questions about the current state of embodiment—involving mediated communication and the moral quality of human interaction, the significance of reflexive engagements with bodily identities, and competing conceptions of how bodies and bodily acts can be considered sacred—shows how body matters reach beyond the boundaries of enfleshed persons. Embodied individuals are always situated within a wider social and material environment that they both shape and are shaped by. The manner in which bodies are conceptualized, experienced, lived, and treated, therefore, provides us with far more than a limited and localized topic—of interest to only physiologists and others in the biological sciences. Instead, these issues provide key means of approaching social relationships, cultural ideas, technological developments, and historical change.

References

Introduction

Kathy Davis (2003) 'Surgical passing: or why Michael Jackson's nose makes "us" uneasy', *Feminist Theory* 4(1): 73–92.

Sander L. Gilman (2000) *Making the Body Beautiful: A Cultural History of Aesthetic Surgery*. Princeton: Princeton University Press.

Chapter 1: Natural bodies or social bodies?

Donella H. Meadows, Dennis L. Meadows, Jorgen Randers, and William W. Behrens III (1972) *Limits to Growth*. New York: New American Library.

Erving Goffman (1959) *The Presentation of Self in Everyday Life*. Harmondsworth: Penguin.

Ted Benton (1996) *Natural Relations: Ecology, Animal Rights and Social Justice*. London: Verso.

Rebecca Gowland and Tim Thompson (2013) *Human Identity and Identification*. Cambridge: Cambridge University Press.

Elizabeth Grosz (2011) *Becoming Undone: Darwinian Reflections on Life, Politics, and Art*. Durham: Duke University Press.

Marcel Mauss (1973 [1934]) 'Techniques of the body', *Economy and Society* 2: 70–88.

Chapter 2: Sexed bodies

John Gray (1992) *Men are from Mars, Women are from Venus*. London: Harper Collins.

Thomas Laqueur (1990) *Making Sex: Body and Gender from the Greeks to Freud*. Cambridge, MA: Harvard University Press.
Richard Dawkins (1976) *The Selfish Gene*. London: Paladin.
Simone de Beauvoir (1949) *The Second Sex*. London: Everyman.
Raewyn Connell (1987) *Gender and Power*. Oxford: Polity.
Raewyn Connell (2005) *Masculinities*, 2nd edn. Oxford: Polity.
Jennifer Hargreaves (1987) 'Victorian familialism and the formative years of female sport', in John Mangan and Roberta Park (eds) *From 'Fair Sex' to Feminism: Sport and the Socialization of Women in the Industrial and Post-Industrial Eras*. London: Cass, p.134.
Esther Newton (1979) *Mother Camp*. Chicago: Chicago University Press.
Judith Butler (1990) *Gender Trouble: Feminism and the Subversion of Identity*. London: Routledge.

Chapter 3: Educating bodies

Basil Bernstein (2000) *Pedagogy, Symbolic Control and Identity*, 2nd edn. New York: Lanham.
Marcel Mauss (1973 [1934]) 'Techniques of the body', *Economy and Society* 2: 70–88.
David Sudnow (2002) *Ways of the Hand: A Rewritten Account*. Harvard, MA: MIT Press.
Maurice Merleau-Ponty (1962) *Phenomenology of Perception*. London: Routledge.
Cristina Grasseni (2007) 'Good looking: learning to be a cattle breeder', in Cristina Grasseni (ed.) *Skilled Visions: Between Apprenticeship and Standards*. Oxford: Berghahn.
Rane Willerslev (2007) *Soul Hunters: Hunting, Animism and Personhood among the Siberian Yukaghirs*. Berkeley, CA: University of California Press.
Barry Saunders (2007) 'CT suite: visual apprenticeship in the age of the mechanical viewbox', in Cristina Grasseni (ed.) *Skilled Visions: Between Apprenticeship and Standards*. Oxford: Berghahn.
Andreas Roepstorff (2007) 'Navigating the brainscape: when knowing becomes seeing', in Cristina Grasseni (ed.) *Skilled Visions: Between Apprenticeship and Standards*. Oxford: Berghahn.
Loic Wacquant (2004) *Body & Soul: Notebooks of an Apprentice*. Oxford: Oxford University Press, pp.66, 87.
Tony Watling (2005) 'Experiencing Alpha', *Journal of Contemporary Religion* 20(1): 91–108.

Tim Ingold (2000) *The Perception of the Environment: Essays on Livelihood, Dwelling and Skill.* London: Routledge, p.409.

John Evans, Emma Rich, Brian Davies, and Rebecca Allwood (2008) *Education, Disordered Eating and Obesity Discourse.* London: Routledge.

Chapter 4: Governing bodies

Bryan S. Turner (2008) *The Body and Society*, 3rd edn. London: Sage.

Thomas Hobbes (1651) *The Leviathan.* Oxford: Oxford University Press.

Michel Foucault (1978) *The History of Sexuality.* Vol. 1: *An Introduction.* Harmondsworth: Penguin, p.140.

Michel Foucault (1975) *Discipline and Punish.* Harmondsworth: Penguin.

Richard Overy (2010) *The Morbid Age: Britain and the Crisis of Civilization 1919–1939.* Harmondsworth: Penguin.

Michel Foucault (1980) 'Body/power', in C. Gordon (ed.) *Michel Foucault: Power/Knowledge.* Brighton: Harvester, p.57.

Ann Davis (1997) 'The body as password', *Wired* July, available from: <http://archive.wired.com/wired/archive/5.07/biometrics_pr.html> accessed 10 November 2014.

Nikolas Rose (2007) *The Politics of Life Itself.* Princeton: Princeton University Press.

Kaushik Sunder Rajan (2006) *Bio Capital: The Constitution of Post-genomic Life.* Durham: Duke University Press.

Tom Shakespeare (1998) 'Choices and rights: eugenics, genetics and disability equality', *Disability and Society* 13(5): 665–81.

Troy Duster (2015) 'A post-genomic surprise: the molecular reinscription of race in science, law and medicine', *British Journal of Sociology* 66(1): 1–27.

Melinda Cooper (2008) *Life as Surplus: Biotechnology and Capitalism in the Neoliberal Era.* Washington, DC: University of Washington Press.

Chapter 5: Bodies as commodities

Rhys Blakeley (2014) 'Past it at 32: the geeks of Silicon Valley stay fresh-faced with Botox', *The Times*, 29 March.

Barbara Ehrenreich (1989) *Fear of Falling: The Inner Life of the Middle Class.* New York: HarperPerennial.

Ruth Holliday and Joanna Elfving-Hwang (2012) 'Gender, globalization and aesthetic surgery in South Korea', *Body and Society* 18: 58–81.
Pierre Bourdieu (1984) *Distinction: A Social Critique of the Judgment of Taste*. London: Routledge.
Catherine Waldby (2002) 'Stem cells, tissue cultures and the production of biovalue', *Health* 6(3): 305–23.
Joseph Dumit (2012) *Drugs for Life*. Durham: Duke University Press.
Kaushik Sunder Rajan (2010) 'The experimental machinery of global clinical trials', in Aihwa Ong and Nancy Chen (eds) *Asian Biotech*. Durham: Duke University Press.
Nancy Scheper-Hughes (2011) 'Mr Tati's holiday and Joao's safari: seeing the world through transplant tourism', *Body & Society* 17(2–3): 55–92.
Susanne Lundin (2008) 'The valuable body: organ trafficking in Eastern Europe', *Baltic Worlds* 1(1): 6–9.
David Harvey (2009) *A Companion to Marx's Capital*. London: Verso.
Penelope McRedmond (2010) 'Defining organised crime in the context of human trafficking', in Gillian Wylie and Penelope McRedmond (eds) *Human Trafficking in Europe*. London: Palgrave Macmillan.
Karl Marx (1972) 'The life-destroying toil of slaves', in Saul Padover (ed.) *The Karl Marx Library*. Vol. II: *On America and the Civil War*. New York: McGraw-Hill, p.21.
Klaus Hoeyer, Sniff Nexoe, Mette Hartlev, and Lene Koch (2009) 'Embryonic entitlements: stem cell patenting and the co-production of commodities and personhood', *Body & Society* 15(1): 1–24.
Aslihan Sanal (2011) *New Organs Within Us: Transplants and the Moral Economy*. Durham: Duke University Press.
David McLellan (1977) *Karl Marx: Selected Writings*. Oxford: Oxford University Press, pp.192–4.

Chapter 6: Bodies matter: dilemmas and controversies

Johan Goudsblom (1992) *Fire and Civilization*. London: Allen Lane.
Vince Miller (2012) 'A crisis of presence: on-line culture and being in the world', *Space and Polity* 16(3): 265–85.
Norbert Elias (2000 [1939]) *The Civilizing Process*. Oxford: Blackwell.
Birgit Meyer (2013) 'Mediation and immediacy: sensational forms, semiotic ideologies and the question of the medium', in J. Boddy

and M. Lambek (eds) *A Companion to the Anthropology of Religion*. Oxford: Wiley-Blackwell.

Chris Shilling (2005) 'Embodiment, emotions and the foundations of social order: Durkheim's enduring contribution', in Jeffrey Alexander and Philip Smith (eds) *The Cambridge Companion to Emile Durkheim*. Cambridge: Cambridge University Press, pp.211–38.

Georgio Agamben (1998) *Homo Sacer: Sovereign Power and Bare Life*. Stanford: Stanford University Press.

Further reading

Introduction

On philosophical approaches to the body, see Donn Welton (1999) (ed.) *The Body: Classic and Contemporary Readings*. Oxford: Blackwell.

Chapter 1: Natural bodies or social bodies?

On the rise of the field 'body studies', see Bryan S. Turner (1991) 'Recent developments in the theory of the body', in Mike Featherstone, Mike Hepworth, and Bryan S. Turner (eds) *The Body: Social Process and Cultural Theory*. London: Sage. See also Chris Shilling (1993) *The Body and Social Theory*. London: Sage.

On the relevance of pragmatism for studying embodiment, see Chris Shilling (2008) *Changing Bodies: Habit, Crisis and Creativity*. London: Sage.

Chapter 2: Sexed bodies

On Gustave Le Bon, see Steven Jay Gould (1981) *The Mismeasure of Man*. Harmondsworth: Penguin.

On endocrinology, psychological, and other theories of embodied sex difference, see Jonanna Meyerowitz (2002) *How Sex Changed*. Cambridge, MA: Harvard University Press.

Chapter 3: Educating bodies

On body pedagogics, see Chris Shilling and Philip A. Mellor (2007) 'Cultures of embodied experience: technology, religion and body pedagogics', *The Sociological Review* 55(3): 531–49.

On John Dewey, and noetic and anoetic knowledge, see Jim Garrison (2015) 'Dewey's aesthetics of body-mind functioning', in Alfonsina Scarinzi (ed.) *Aesthetics and the Embodied Mind: Beyond Art Theory and the Cartesian Mind-Body Dichotomy*. Dordrecht: Springer.

Chapter 4: Governing bodies

On ethical issues and bodily commodification, see Stephen Wilkinson (2003) *Bodies for Sale: Ethics and Exploitation in the Human Body Trade*. London: Routledge.

Chapter 5: Bodies as commodities

See US Department of State Trafficking in Persons Report 2015, available from: <http://www.state.gov/j/tip/rls/tiprpt/2014/?utm_source=NEW+RESOURCE:+Trafficking+in+Persons+R>.

On Kant, philosophy, and ethical issues regarding transplant surgery, see David Petechuk (2006) *Organ Transplantation*. London: Greenwood Press.

Chapter 6: Bodies matter: dilemmas and controversies

On the secular and religious conceptions of the sacred, and their role in contemporary conflict, see Philip A. Mellor and Chris Shilling (2014) *Sociology of the Sacred: Religion, Embodiment and Social Change*. London: Sage.

Index

9/11 attack 12, 70, 78

A

abortion 8, 108
Agamben, Giorgio 69, 79
Alpha Course 48, 54, 104
Ancient Greece 1, 43, 61, 69
anoetic/noetic knowledge 56–7
Aristotle 61

B

bare life 69, 77, 79
Bentham, Jeremy 65–6
Bentham, Samuel 66
Benton, Ted 16, 18
Bernstein, Basil 45
bio value 84
bioarchaeology 18–19
biological
  citizenship 76–7, 106
  reproduction 24, 33
  sciences 1, 4, 8, 16, 18–19, 21–2, 27, 36, 108
biometric data 12, 73
biopolitics 62, 64, 66, 68–70, 78–9
biotechnology 68, 81, 84–7
bioterrorism 78–9
body
  ageing 9–10, 18, 23, 35, 76
  as a commodity 4–5, 80–96, 106–7
  educated/educating 4–5, 41–59
  elusiveness of 22–3
  enslaving the 81, 89–93, 96
  gendered 29–36, 38–40, 42
  governed 4, 60–79, 106–7
  inferior female 27, 29, 31, 33
  medicalized 84–7, 93
  modification 2, 12
  natural 5, 7–23, 27–31, 39–41, 102, 106
  pedagogics 42–5, 48, 50–9, 104
  as a project 13–16, 40, 84
  sacred 6, 98, 106–8
  sexed 4, 24–41
  social 7–23
  studies xiv–xv, 7, 12, 22
  surveillance of 12, 60, 64, 66–8, 70–5, 78
  techniques of 43, 45–9, 51–2, 54–6, 58, 82, 103
bodywork 82–3
Bourdieu, Pierre 83
boxing 51–3
Britain/UK 18, 25, 29, 40, 43–4, 67–8, 71–2, 87, 108
  Victorian 25, 33–4, 44

Burt, Cyril 68
Bush, George, W. 12, 70, 78
Butler, Judith 36–41

C

China 39, 68, 73, 90–1
Christians/Christianity 3, 12, 23, 27, 30, 55, 107–8
 charismatic 48, 103
 evangelical 54, 57
 muscular 44
 Pentecostalism 103–4
closed circuit television (CCTV) 71–2
communication 10–11, 16–17, 24, 50, 57, 94, 98, 103
 digitally mediated 1, 12, 98–101, 108
 moral quality of 98–101
Comte, Auguste 22
Confucianism 2
Connell, Raewyn 32–4, 41
consumer culture 8–10, 58, 69
Cooper, Melinda 78
cosmetic surgery 2, 4, 10, 12–13, 82–3
craniometry 28

D

Daoism 2
Darwin, Charles 19, 28, 67
Davis, Ann 73
Dawkins, Richard 29
de Beauvoir, Simone 31–2, 41
Descartes, Rene 2–3
Dewey, John 3, 22, 56, 102
diet/ing 6–7, 13, 16, 18–19, 33, 44, 52–3, 74, 76, 82, 103, 107
disability 7, 11, 68–9, 76–7, 108
Dumit, Joseph 85
Durkheim, Emile 106
Duster, Troy 77

E

eating disorders 34, 58
Ehrenreich, Barbara 82
Elias, Norbert 61, 101
emotion 17, 21, 52, 54, 82, 85, 103
endocrinology 35–6
Enlightenment 28
ethnicity xiv, 2, 7, 91, 106
eugenics 66–8, 76–7
Europe 9–10, 46, 78, 83, 88, 90, 104
 European Union 73, 91
euthanasia 10
Evans, John 58

F

feminism 8, 19, 23, 30–2, 36, 40, 42
forced labour 5, 90–3, 96
Foucault, Michel 3, 62, 64, 66, 69, 72, 77
Frieden, Betty 30

G

Galen, Claudius 26
Galton, Francis 67
genetics 4, 7, 19–21, 29, 36, 75–8, 84–6
George, David Lloyd 68
Germany 19, 68, 72
Goffman, Erving 11, 99
Gowland, Rebecca 18
Grasseni, Cristina 49
Greer, Germaine 30
Grosz, Elizabeth 19

H

Hargreaves, Jennifer 33
Harvey, David 89
health and fitness 1, 7–8, 10, 13, 18–19, 21, 25, 51, 58, 64, 67–8, 74, 76–9, 85, 88–9, 94, 105–7
Hobbes, Thomas 27, 61

Hochschild, Arlene 34
Hoeyer, Klaus 94
human genome 36, 75, 86
humanities xiv, 1, 7–8, 16
Huxley, Julian 68

I

Ingold, Tim 57
in vitro fertilization 4
Internet, the 17, 70, 73, 98, 100, 103
Islam 55, 74, 104, 107–8

J

James, William 22
Jefferson, Thomas 85
Joseph, Jacques 2

K

Kant, Immanuel 95
Keynes, John Maynard 68

L

Laqueur, Thomas 26
Le Bon, Gustave 28
Levinas, Emmanuel 101
Locke, John 27
Lundin, Susanne 88

M

McRedmond, Penelope 90
Mahmood, Saba 104
Marx, Karl 80, 92, 95–6
Mauss, Marcel 43, 45, 55
Mead, George Herbert 22
media *see* communication
medical tourism 89
medicine 4, 8–9, 21, 29, 49–51, 56, 69, 81, 88–9, 96, 101 *see also* body, governed; body, medicalized
medieval era 23, 30, 62–5, 79, 101
Merleau-Ponty, Maurice 3, 48
Middle East 88, 91
migrant labour 90–1
Miller, Vince 99–100
Millett, Kate 30
mind/body dualism 1–3, 42–3, 56, 95, 97, 102, 106
mirror neurones 34–5

N

neural implants 1
neuroscience 20, 34–5, 75
Newton, Esther 36
Nietzsche, Friedrich 3

O

Oakley, Anne 30
obesity 1, 20, 58, 85
one sex/one flesh model 26–8, 30, 35
Östman, Leif 47

P

panopticon 65–6, 70
patients in waiting 76–7
performativity 38–9
pharmacogenomics 85
Plato 61
pornography 8, 32, 73, 92
presentation of self 11
primitive accumulation 89–90, 96
prison reform 65–6
prostitution 8, 91–2

Q

quantified-self 74–5

R

'race' 7, 25, 67, 83
Rajan, Kaushik Sunder 76
reflexivity 5, 13, 56, 73, 102–5, 108
religion xiv, 48, 55, 57, 102–5
Roe, John 2
Rose, Nikolas 75, 77
Rousseau, Jean-Jacques 61

S

Sanal, Aslihan 95
Sanger, Margaret 67
scarification 2, 7
Scheper-Hughes, Nancy 87
science 4, 12, 18, 27–30, 35, 75–8, 81, 84, 101
scientific management 12, 66
self-identity 11, 13, 18, 105
sex
  'differences' 23–6, 30–6, 38, 40
  industry 5, 81, 92
  inequalities 7–9, 28–30, 32
  see also body, sexed
Shakespeare, Tom 77
Shelley, Louise 92
sight 18, 48–52
slavery 2, 5, 18, 61, 80–1, 89–93, 96
Smith, Adam 101
social class 2, 25, 44, 46, 80, 82–3
  inequalities xiv, 4, 9, 26, 28–30, 33, 78, 87, 92
social media *see* communication
social sciences xiv, 1, 4, 7–8, 16, 18–19, 21–2, 35–6
sociobiology 25, 29, 36
sociology xiv, 16, 22
stem cell research 20
Stopes, Marie 67
Sudnow, David 48
surveillance 12, 60, 70–5, 78
  productive 64, 66, 68
  see also body, governed

T

Taylor, Frederick Winslow 66
tattooing 2, 7, 13, 15, 91
technology 1, 4, 8, 12, 16–18, 23, 74–6, 78, 82, 97–9, 101, 108
  see also biotechnology; communication
terrorists 12, 70, 72, 78–9, 92, 108
torture 62–3, 65, 71, 101
trafficking
  body parts/organs 81, 87–9, 93
  human 5, 90–2
transgenderism 38–9
transplant
  surgery 4, 12, 89, 95
  tourism/trafficking 87–8
Turner, Bryan S. 60–1, 72

U

Ukraine 88, 92
USA 10, 12–13, 18, 29, 40, 44, 66–73, 77–8, 82–3, 85–7, 91, 95, 101

V

Venter, Craig 86
violence 3, 8, 46, 61, 64, 73–4, 90, 108
von Helmholtz, Hermann 66

W

Wacquant, Loic 51–3
Waldby, Catherine 84
War on Terror 12, 70–1, 78
Watling, Tony 54
Willerslev, Rane 49–50
World Health Organisation (WHO) 94
World War I 44, 67
World War II 2, 68

FEMINISM

A Very Short Introduction

Margaret Walters

How much have women's lives really changed? What are we to make of the now commonplace insistence that feminism deprives men of their rights and dignities? And how does one tackle the issue of female emancipation in different cultural and economic environments? This book provides an historical account of feminism, exploring its earliest roots as well as key issues including voting rights, the liberation of the sixties, and its relevance today. Margaret Walters touches on the difficulties and inequities that women still face more than forty years after the 'new wave' of 1960s feminism. She provides an analysis of the current situation of women across the globe, from Europe and the United States to Third World countries.

{No reviews}

www.oup.com/vsi

SOCIAL AND CULTURAL ANTHROPOLOGY

A Very Short Introduction

John Monaghan and Peter Just

This *Very Short Introduction* to Social and Cultural Anthropology combines an accessible account of some of the disciplines guiding principles and methodology with abundant examples and illustrations of anthropologists at work. Peter Just and John Monaghan discuss anthropology's most important contributions to modern thought from its investigation of culture as a distinctively human characteristic to its doctrine of cultural relativism. How has social and cultural anthropology advanced our understanding of human society and culture? And what is its likely future?

{No reviews}

www.oup.com/vsi

SOCIOLOGY

A Very Short Introduction

Steve Bruce

Drawing on studies of social class, crime and deviance, work in bureaucracies, and changes in religious and political organizations, this *Very Short Introduction* explores the tension between the individual's role in society and society's role in shaping the individual, and demonstrates the value of sociology as a perspective for understanding the modern world.

{No reviews}

www.oup.com/vsi

SOCIAL MEDIA
Very Short Introduction

Join our community

www.oup.com/vsi

- Join us online at the official Very Short Introductions **Facebook** page.
- Access the thoughts and musings of our authors with our online **blog**.
- Sign up for our monthly **e-newsletter** to receive information on all new titles publishing that month.
- Browse the full range of Very Short Introductions online.
- Read **extracts** from the Introductions for free.
- Visit our library of **Reading Guides**. These guides, written by our expert authors will help you to question again, why you think what you think.
- If you are a teacher or lecturer you can order inspection copies quickly and simply via our website.